The Speech of Monkeys

by R. L. Garner

PREFACE

I desire here to express my gratitude to The New Review, The North American Review, The Cosmopolitan, The Forum, and many of the leading journals of America, for the use of their valuable and popular pages through which my work has been given to the public. To the press, English and American, I gladly pay my tribute of thanks for the liberal discussion, candid criticism, and kind consideration which they have bestowed upon my efforts to solve the great problem of speech.

In contributing to Science this mite, I do not mean to intimate that my task has been completed, for I am aware that I have only begun to explore the field through which we may hope to pass beyond the confines of our own realm and invade the lower spheres of life.

This volume is intended as a record of my work, and a voluntary report of my progress, to let the world know with what results my labours have been rewarded, and with the hope that it may be the means of inducing others to pursue like investigations.

In prosecuting my studies I have had no precedents to guide me, no literature to consult, and no landmarks by which to steer my course. I have, therefore, been compelled to find my own means, suggest my own experiments, and solve my own problems. Not a line on this subject is to be found in all the literature of the world, and yet the results which I have obtained have far surpassed my highest hopes. Considering the difficulties under which I have been compelled to work, I have been rewarded with results for which I dared not hope, and this inspires me to believe that my success will meet my highest wishes when I am placed in touch with such subjects as I expect to find in the forests of Tropical Africa.

Only a few of my experiments are recorded in this volume, but as they illustrate my methods and set forth the results, they will serve to show, in a measure, the scope of my work.

In the latter part of this work will be found a definition of the word Speech as I have used it, and the deductions which I have made from my experiments. I have not ventured into any extreme theories, either to confirm or controvert the opinions of others, but simply commit to the world these initial facts, and the working hypotheses upon which I have proceeded to obtain them.

In Chapter XXI. I have mentioned the particular characteristics which mark the sound of monkeys as speech, and distinguish them from mere automatic sounds.

With all the gravity of sincere conviction I commit this volume to the friends of Science as the first contribution upon this subject.

R. L. GARNER.

NEW YORK, June 1, 1892.

CONTENTS

CHAPTER XXII.

CHAPTER XXIII.

CHAPTER XXIV.

CHAPTER XXV.

CHAPTER XXVI.

THE SPEECH AND REASON OF DOMESTIC ANIMALS.

* * * * *

THE SPEECH OF MONKEYS

CHAPTER I.

Early Impressions--First Observations of Monkeys--First Efforts to Learn their Speech--Barriers--The Phonograph Used--A Visit to Jokes--My Efforts to Speak to Him--The Sound of Alarm inspires Terror.

From childhood, I have believed that all kinds of animals have some mode of speech by which they could talk among their own kind, and have often wondered why man had never tried to learn it. I often wondered how it occurred to man to whistle to a horse or dog instead of using some sound more like their own; and even yet I am at a loss to know how such a sound has ever become a fixed means of calling these animals. I was not alone in my belief that all animals had some way to make known to others some certain things; but to my mind the means had never been well defined.

[Sidenote: FIRST OBSERVATIONS OF MONKEYS]

About eight years ago, in the Cincinnati Zoological Garden, I was deeply impressed by the conduct of a number of monkeys occupying a cage with a huge, savage mandril, which they seemed very much to fear and dislike. By means of a wall, the cage was divided into two compartments, through which was a small doorway, just large enough to allow the occupants of the cage to pass from one room to the other. The inner compartment of the cage was used for their winter quarters and sleeping apartments; the outer, consisting simply of a well-constructed iron cage, was intended for exercise and summer occupancy. Every movement of this mandril seemed to be closely watched by the monkeys that were in a position to see him, and instantly reported to the others in the adjoining compartment. I watched them for hours, and felt assured that they had a form of speech by means of which they communicated with each other. During the time I remained, I discovered that a certain sound would invariably cause them to act in a certain way, and, in the course of my visit, I discovered that I could myself tell, by the sounds the monkeys would make, just what the mandril was doing--that is, I could tell whether he was asleep or whether he was moving about in his cage. Having interpreted one or two of these sounds, I felt inspired with the belief that I could learn them, and felt that the "key to the secret chamber" was within my grasp.

I regarded the task of learning the speech of a monkey as very much the same as learning that of some strange race of mankind, more difficult in the degree of its inferiority, but less in volume.

Year by year, as new ideas were revealed to me, new barriers arose, and I began to realise how great a task was mine. One difficulty was to utter the sounds I heard, another was to recall them, and yet another to translate them. But impelled by an inordinate hope and not discouraged by poor success, I continued my studies, as best I could, in the Gardens of New York, Philadelphia, Cincinnati and Chicago, and with such specimens as I could find from time to time with travelling shows, hand-organs, aboard some ship, or kept as a family pet. I must acknowledge my debt of gratitude to all these little creatures who have aided me in the study of their native tongue.

[Sidenote: ACTING AS INTERPRETER]

Having contended for some years with the difficulties mentioned, a new idea dawned upon me, and, after maturely considering it, I felt assured of ultimate success. I went to Washington, and proposed the novel experiment of acting as interpreter between two monkeys. Of course this first evoked from the great fathers of science a smile of incredulity; but when I explained the means by which I expected to accomplish this, a shadow of seriousness came over the faces of those dignitaries to whom I first proposed the novel feat. I procured a phonograph upon which to record the sounds of the monkeys. I separated two monkeys which had occupied the same cage together for some time, and placed them in separate rooms of the building where they could not see or hear each other. I then arranged the phonograph near the cage of the female, and by various means induced her to utter a few sounds, which were recorded on the cylinder of the phonograph. The machine was then placed near the cage containing the male, and the record repeated to him and his conduct closely studied. He gave evident signs of recognising the sounds, and at once began a search for the mysterious monkey doing the talking. His perplexity at this strange affair cannot well be described. The familiar voice of his mate would induce him to approach, but that squeaking, chattering horn was a feature which he could not comprehend. He traced the sounds, however, to the horn from which they came, and, failing to find his mate, thrust his arm into the horn quite up to his shoulder, then withdrew it, and peeped into it again and again. The expressions of his face were indeed a study. I then secured a few sounds of his voice and delivered them to the female, who showed some signs of interest, but the record was very imperfect and her manner seemed quite indifferent. In this experiment, for the first time in the history of language, was the Simian speech reduced to record; and while the results were not fully up to my hopes, they served to inspire me to further efforts to find the fountain-head from which flows out the great river of human speech. Having satisfied myself that each one recognised the sound made by the other when delivered through the phonograph, I felt rewarded for my labour and assured of the possibility of learning the language of monkeys. The faith of others was strengthened also, and while this experiment was very crude and imperfect, it served to convince me that my opinions were correct as to the speech of these animals.

[Sidenote: RECORDS OF SOUNDS]

In this case I noticed the defects which occurred in my work and provided against them, as well as I could, for the future. Soon after this I went to Chicago and Cincinnati, where I made a number of records of the sounds of a great number of monkeys, and among others I secured a splendid record of the two chimpanzees contained in the Cincinnati collection, which I brought home with me for study. The records that I made of various specimens of the Simian race I repeated to myself over and over, until I became familiar with them, and learned to imitate a few of them, mostly by the use of mechanical devices. After having accomplished this I returned to Chicago, and went at once to visit a small Capuchin monkey whose record had been my chief study. Standing near his cage, I imitated a sound which I had translated "milk," but from many tests I concluded it meant "food," which opinion has been somewhat modified by many later experiments which led me to believe that he uses it in a still wider sense. It is difficult to find any formula of human speech equivalent to it. While the Capuchin uses it relating to food and sometimes to drink, I was unable to detect any difference in the sounds. He also seemed to connect the same sound to every kindly office done him, and to use it as a kind of "Shibboleth." More recently, however, I have detected in the sound slight changes of inflection under different conditions, until I am now led to believe that the meaning of the word depends somewhat, if not wholly, on its modulation. The phonetic effect is rich and rather flute-like, and the word resembles somewhat the word "who." Its dominant is a pure vocal "u," sounded like "oo" in "too," which has a faint initial "wh," both elements of which are sounded, and the word ends with a vanishing "w." The literal formula by which I would represent it is "wh-oo-w." The word which I have translated "drink" begins with a faint guttural "ch," and glides through a sound resembling the French diphthong "eu," and ends with a slight "y" sound as in "ye."

So far I have found no trace of the English vowels "a," "i," or "o," unless it be in the sound emitted under stress of great alarm or in case of assault, in which I find a close resemblance to the vowel "i," short as in "it."

[Sidenote: FIELD OF OPERATIONS EXTENDED]

After having acquired a sound or two, I extended my field of operations and began to try my skill as a Simian linguist on every specimen with which I came in contact.

In Charleston, a gentleman owns a fine specimen of the brown Cebus whose name is Jokes. He is naturally shy of strangers, but on my first visit to him I addressed him in his native tongue, and he really seemed to regard me very kindly; he would eat from my hand and allow me to caress him through the bars of his cage. He eyed me with evident curiosity, but invariably responded to the word which I uttered in his own language. On my third visit to him I determined to try the effect of the peculiar sound of "alarm" or "assault" which I had learned from one of this species; but I cannot very well represent it in letters. While he was eating from my hand, I gave this peculiar piercing note, and he instantly sprang to a perch in the top of his cage, thence in and out of his sleeping apartment with great speed, and almost wild with fear.

[Sidenote: HARSH MEANS RESORTED TO]

As I repeated the sound his fears seemed to increase, until from a mere sense of compassion I desisted. No amount of coaxing would induce him to return to me or to accept any offer of peace which I could make. I retired to a distance of about twenty feet from his cage, and his master induced him to descend from the perch, which he did, with the greatest reluctance and suspicion. I gave the sound again from where I stood, and it produced almost the same results as before. The monkey gave out a singular sound in response to my efforts to appease him, but refused to become reconciled. After the lapse of eight or ten days, I had not been able to reinstate myself in his good graces, or to induce him to accept anything whatever from me. At this juncture I resorted to harsher means of bringing him to terms, and began to threaten him with a rod. At first he resented this, but soon yielded and came down merely from fear. He would place the side of his head on the floor, put out his tongue, and utter a very plaintive sound having a slight interrogative inflection. At first this act quite defied interpretation; but during the same period I was visiting a little monkey called Jack. For strangers, we were quite good friends, and Jack allowed me many liberties which the family assured me he had uniformly refused to others. On one of my visits he displayed his temper, and made an attack upon me because I refused to let go of a saucer from which I was feeding him with some milk. I jerked him up by the chain and slapped him sharply, whereupon he instantly laid the side of his head on the floor, put out his tongue, and made just such a sound as Jokes had made a number of times before. It occurred to me that it was a sign of surrender, and many subsequent tests have confirmed this opinion. Mrs. M. French Sheldon, in her journey through East Africa, shot a small monkey in a forest near Lake Charla. She described to me how the little fellow stood high up in a tree and chattered to her in his sharp, musical voice, until at the crack of her gun he fell mortally wounded. When he was laid dying at her feet, he turned his bright little eyes pleadingly upon her as if to ask for pity. Touched by his appeal, she took the little creature in her arms to try to soothe him. Again and again he would touch his tongue to her hand as if kissing it, and seemed to wish in the hour of death to be caressed, even by the hand that slew him, and which had taken from him without reward that life which could be of no value except to spend in the wild forest where his kindred monkeys live.

[Sidenote: MODE OF EXPRESSING SUBMISSION]

This peculiar mode of expressing submission seems to be very widely used, and from her description of the actions of that monkey, his conduct must have been identical with that of the Cebus; and to my mind may justly be interpreted to mean, "Pity me, I will not harm you." I have recently learned that a Scotch naturalist, commenting on my description of this act and its meaning, quite agrees with me, and states that he has observed the same thing in other species of monkeys.

CHAPTER II.

The Reconciliation--The Acquaintance of Jennie--The Salutation--The Words for Food and Drink--Little Banquo, Dago, McGinty, and others.

During a period of many weeks I visited Jokes almost daily, but after the lapse of more than two months I had not won him back nor quieted his suspicions against me. On my approach, he would manifest great fear and go through the act of humiliation described above. I observed that he entertained an intense hatred for a negro boy on the place, who teased and vexed him on all occasions. I had the boy come near the cage, and Jokes fairly raved with anger. I took a stick and pretended to beat the boy, and this delighted Jokes very greatly. I held the boy near enough to the cage to allow the monkey to scratch and pull his clothes, and this would fill his little Simian soul with joy. I would then release the boy, and to the evident pleasure of Jokes I would drive him away by throwing wads of paper at him. I repeated this a number of times, and by such means we again became the best of friends. After each encounter with the boy, he would come up to the bars, touch my hand with his tongue, chatter and play with my fingers, and show every sign of confidence and friendship. He always warned me of the approach of any one, and his conduct towards them was largely governed by my own. He never failed, after this, to salute me with the sound described in the first chapter. About the same time I paid a few visits to another little monkey of the same species, named "Jennie." Her master had warned me in advance that she was not well disposed towards strangers. At my request, he had her chained in a small side yard which he forbade any of the family entering. When I approached the little lady for the first time, I gave her the usual salutation, which she responded to, and seemed to understand. I unceremoniously sat down by her side and fed her from my hands. She eyed me with evident interest and curiosity, while I studied her every act and expression. During the process of this mutual investigation, a negro girl who lived with the family, overcome by curiosity, stealthily came into the yard and came up within a few feet of us. I determined to sacrifice this girl upon the altar of science, so I arose and placed her between the monkey and myself, and vigorously sounded the alarm or menace. "Jennie" flew into a fury, while I continued to sound the alarm and at the same time pretended to attack the girl with a club and some paper wads, thus causing the monkey to believe that the girl had uttered the alarm and made the assault. I then drove the girl from the yard with a great show of violence, and for days afterwards she could not feed or approach the little Simian. This confirmed my opinion of the meaning of the sound, which can be fairly imitated by placing the back of the hand gently on the mouth and kissing it with great force, prolonging the sound for some seconds. This imitation, however, is indifferent, and its quality is especially noticeable when analysed on the phonograph. The pitch corresponds to the highest "F" sharp on the piano, while the word "food" is four octaves lower and the word "drink" three.

[Sidenote: THE GARDEN IN CINCINNATI]

On one occasion I visited the Garden in Cincinnati, and found in a cage a small Capuchin, to whom I gave the name of Banquo. It was near night and the visitors had left the house, and the little monkey, worried out by the day's annoyance from visitors, sat quietly in the back of his cage as though he was glad another day was done. I approached the cage and uttered the sound which I have described and translated "drink." My first effort caught his attention and caused him to turn and look at me. He then arose and answered me with the same word, and came at once to the front of the cage. He looked at me as if in doubt, and I repeated the word. He responded with the same and turned to a small pan in his cage, which he took up and placed near the door through which the keeper usually passed his food, returned to me, and uttered the word again. I asked the keeper for some milk, which he did not have, but brought me some water instead. The efforts of my little Simian friend to secure the glass were very earnest, and his pleading manner and tone assured me of his extreme thirst. I allowed him to dip his hand into the glass, and he would then lick the water from his fingers and reach again. I kept the glass out of reach of his hand, and he would repeat the sound earnestly and look at me beseechingly, as if to say, "Please give me some more." I was thus convinced that the word which I had translated "milk" must also mean "water," and from this and other tests I at last determined that it meant "drink" in its broad sense, and possibly "thirst." It evidently expressed his desire for something with which to allay his thirst. The sound is very difficult to imitate, and quite impossible to write exactly.

[Sidenote: IMITATING SOUNDS]

On one of my visits to the Chicago Garden, I stood with my side to a cage containing a small Capuchin and gave the sound which I had translated "milk." It caused him to turn and look at me, and on repeating the sound a few times, he answered me very distinctly with the same, picking up the pan from which he usually drank; and as I repeated the word, he brought the pan to the front of the cage, set it down, and came up to the bars and uttered the word distinctly. I had not shown him any milk or any kind of food, but the man in charge, at my request, brought me some milk, which I gave to him. He drank it with great delight, then looked at me and held up his pan, repeating the sound. I am quite sure that he used the same sound each time that he wanted milk. During this same visit, I tried many experiments with the word which I am now convinced means "food" or "hunger." And I was led to the belief that he used the same word for apple, carrot, bread and banana; but a few later experiments have led me to modify this view in a measure, since the phonograph shows me slight variations of the sound, and I now think it probable that these faint inflections may possibly indicate a difference in the kinds of food he has in mind. However, they usually recognise this sound, even when poorly imitated. I am impressed with the firm belief that in this word I have found the clue to the great secret of speech; and while I have taken only one short step in the direction of its solution, I have pointed out the way which leads to it.

[Sidenote: BROWN CAPUCHINS]

In the fall of 1891, I visited New York for the purpose of experimenting with the monkeys in Central Park. Early one morning I repaired to the monkey-house, and for the first time approached a cage containing five brown Capuchins, whom I saluted with the word which I have translated "food," and which seems to be an "open-sesame" to the hearts of all monkeys of this species. On delivering this word, one of them responded promptly and came to the front of the cage. I repeated it two or three times and the remaining four came to the front, and as I thrust my fingers through the bars of the cage, they took hold of them and began playing with great familiarity and apparent pleasure. They seemed to recognise the sound, and to realise that it had been delivered to them by myself. Whether they regarded me as a great ape, monkey, or some other kind of animal speaking their tongue, I do not know. But they evidently understood the sound, though up to this time I had shown them no food or water. A little later I secured some apples and carrots, and gave them in small bits in response to their continual requests for food, and this further confirmed my belief that I had translated the word correctly. This was gratifying to me in view of the fact that I was accompanied by two gentlemen who had been permitted to witness the experiment, and it was evident to them that the monkeys understood the sound. I placed the phonograph in order and made a record of the sound, which I preserved for study. After an absence of some days, I returned to the Park and went to the monkey-house. They recognised me as I entered the door, notwithstanding there were many visitors present. They began begging me to come to their cage, which I did, and gave them my hand to play with. One of them in particular, whose name is "McGinty," showed every sign of pleasure at my visit; he would play with my fingers, hug them, and caress them in the most affectionate manner. Another occupant of the same cage had shown a disposition to become friendly with me, and on this occasion came bravely to the bars of the cage and showed a desire to share the pleasure of my visit with his little Simian brother. But this was denied him on any terms by "McGinty," who pounced upon him and drove him away, as he also did the other monkeys in the cage in order to monopolise my entire society himself. He refused to allow any other inmate of the cage to receive my caresses or any part of the food that I had brought them. I spent the past winter in Washington and New York, much of the time in company with these little creatures, and have made many novel and curious experiments, some of which have resulted in surprises to myself. [Sidenote: MONKEYS CAN COUNT] Among the facts which I have obtained, I may state that certain monkeys can count three; that they discern values by quantity and by number; that they have favourite colours, and are pleased with some musical sounds. And I shall explain how I arrived at some of these conclusions, in order that I may not be supposed to have merely guessed at them.

CHAPTER III.

Monkeys have favourite Colours--Can distinguish Numbers and Quantity--Music and Art very limited.

[Sidenote: MONKEYS HAVE FAVOURITE COLOURS]

In order to ascertain whether monkeys have any choice of colours or not, I selected some bright candies, balls, marbles, bits of ribbon, &c. I took a piece of pasteboard, and on it placed a few bright-coloured bits of candy, which I offered to a monkey and watched to see whether he would select a certain colour or not. In this experiment I generally used two colours at a time, and changed their places from time to time in order to determine whether he selected the colour by design or accident. After having determined which of two colours he preferred, I substituted a third colour for the one which he cared least for, and continued thus until I exhausted the list of bright colours. By changing the arrangement of the objects a great number of times, it could be ascertained with comparative certainty whether the colour was his preference or not. I find that all monkeys do not select the same colour, nor does the same monkey invariably select the same colour at different times; but I think, as a rule, that bright green is a favourite colour with the Capuchin, and their second choice is white. In a few cases, white seemed to be their preference. I have sometimes used paper wads of various colours, or bits of candy of the same flavour rolled in various coloured papers. They seemed to choose the same colours in selecting their toys. I have sometimes used artificial flowers, and find, as a rule, that they will select a flower having many green leaves about it. It may be that they associate this colour with some green food which they are fond of, and consequently that they are influenced by this in selecting other things. I kept a cup for a monkey to drink milk from, on the sides of which were some brilliant flowers and green leaves, and she would frequently quit drinking the milk to play with the flowers on the cup, and seemed never able to understand why she could not get hold of them. In one test I had a board about two feet long, and laid a few pieces of white and pink candies in four places on it. The monkey took the white from each pile before touching the pink, except in one instance it took the pink piece from one pile. I repeated this test many times. In another test I took a white paper ball in one hand and a pink one in the other, and held out my hands to the monkey, who selected the white one nearly every time, although I changed hands with the balls from time to time. These experiments were mostly confined to the Cebus monkeys, but a few of them were made with Macaques. They seem to be attracted generally by all brilliant colours, but when reduced to a choice between two, such seems to be their tastes.

[Sidenote: CAN DISTINGUISH NUMBERS]

In my efforts to ascertain their mathematical skill, I would take in one hand a little platter containing one nut, or one small bit of something to eat, such as a piece of apple or carrot cut into a small cube. In the other hand I held a small platter, with two or three such articles of the same size and colour, and holding them just out of reach of the monkey and changing them from hand to hand, I observed that the monkey would try to reach the one containing the greater number. He readily discerned which platter contained one and which contained two or three pieces. I was long in doubt whether he distinguished by number or by quantity, and my belief was that it was by quantity only. I first determined that he could tell singular from plural, by making the one piece larger and sometimes of a different shape, and from his choice of these I quite satisfied my own mind that he could distinguish by number. [Sidenote: THE TEST WITH MARBLES] I next set out to find how far in numerals his acquirements reached, and after a great number of indecisive trials I fell upon this simple plan: I took a little square wooden box and made a hole in one side just large enough for the monkey to withdraw his hand with a marble in it. I took three marbles of the same size and colour, and gave them to the monkey to play with. After a time I put the marbles in a box and allowed him to take them out, which he could do by taking out only one at a time. I repeated this several times, so as to impress his mind with the number of marbles in the box. I then concealed one of the marbles and returned two to the box. On taking them out, he evidently missed the absent one, felt in the box, arose, and looked around where he had been sitting. Then he would put his hand into the box again and look at me; but failing to find it, he became reconciled, and began to play with the two. When he had become content with the two, I abstracted one of them, and when he failed to find it he began to search for it, and seemed quite unwilling to proceed without it. He would put the one back into the box and take it out again, as if in hope that it might find the other. I helped him to look for the missing marbles, and, of course, soon found them. When he learned that I could find the lost marbles, he would appeal to me as soon as he missed them, and in several instances he would take his little black fingers and open my lips to see if I had concealed them in my mouth, the place where all monkeys conceal what they wish to keep in safety from other monkeys, who never venture to put their fingers into one another's mouth, and when any article is once lodged in a monkey's mouth it is safe from the reach of all the tribe. I repeated this until I felt quite sure of the ability of my subject to count three, and I then increased the number of marbles to four. When I would abstract one of them, sometimes he seemed to miss it, or at least to be in doubt, but would soon proceed with his play and not worry himself about it; yet he rarely failed to show that he was aware that something was wrong. Whether he missed one from four, or only acted on general principles, I do not know; but that he missed one from three was quite evident.

I may here add that there is a great difference in different specimens, and their tastes vary like those of human beings. The same idea is much clearer to some monkeys than it is to others, and a choice of colours much more definite; but I think that all of them assign to different numbers a difference of value. Some are talkative and others taciturn. I think I may state with safety that the Cebus is the most intelligent and talkative of all the monkeys I have known; that the Old World monkeys, as a group, are more taciturn and less intelligent than the New World monkeys, but I do not mean to include the anthropoid apes in this remark.

[Sidenote: MUSICAL RECORDS ON PHONOGRAPH]

As a test of their taste for music or musical sounds, I took three little bells, which I suspended by three strings, one end of which was tied to a button. The bells were all alike, except that from two of them I had removed the clappers. I dropped the bells through the meshes of the cage about a foot apart, and allowed the monkey to play with them. I soon discovered that he was attracted by the one which contained the clapper. He played with it, and soon became quite absorbed in it. I attracted his attention to another part of the cage with some food, and while he was thus diverted I changed the position of the bells by withdrawing and dropping them through other meshes. On his return he would go to the place he had left, and, of course, get a bell with no clapper in it. He would drop this and take another, until he found the one with the clapper, which showed clearly that the sound was a part of the attraction. I have repeated to monkeys many musical records on the phonograph, but frequently they show no sign of concern, while at other times they display some interest. It may be, however, that music, as we understand it, is somewhat too high for them. Musical sounds seem to attract and afford them pleasure, but they do not appreciate melody or rhythm. As monkeys readily discern the larger of two pieces of food from the smaller, and by the aid of concrete things can count a limited number, I feel justified in saying that they have the first principles of mathematics as dealing with numbers and quantity in a concrete form. Their ability to distinguish colours and their selection thereof, would indicate that they possess the first rudiment of art as dealing with colour. And the fact that they are attracted in a slight degree by musical sounds shows that they possess the germ from which music itself is born. I must not be understood to claim that they possess anything more than the mere germ from which such faculties might have been evolved. I do not think that they have any names for numbers, colours or quantities, nor do I think that they possess an abstract idea of these things, except in the feeblest degree; but as the concrete must have preceded the abstract idea in the development of human reason, it impresses me that these creatures are now in a condition such as man has once passed through in the course of his evolution; and it is not difficult to understand how such feeble faculties may develop into the very highest degree of strength and usefulness by constant use and culture.

[Sidenote: RUDIMENTS OF FACULTIES]

We find in them the rudiments from which all the faculties possessed by man could easily develop, including thought, reason, speech, and the moral and social traits of man. In brief, they appear to have at least the raw material out of which is made the most exalted attributes of man, and I shall not contest with them the right of such possession.

CHAPTER IV.

Pedro's Speech Recorded--Delivered to Puck through the Phonograph--Little Darwin learns a new Word.

[Sidenote: PEDRO THE CAPUCHIN]

In the Washington collection there is a little Capuchin by the name of Pedro. When I first visited this bright little monk he occupied a cage in common with several other monkeys of different kinds. All of them seemed to impose upon little Pedro, and a young spider monkey in the cage found special delight in catching him by the tail and dragging him around the floor of the cage. I interfered on behalf of Pedro, and drove the spider monkey away. On account of this, Pedro soon began to look upon me as his benefactor, and when he would see me he would scream and beg for me to come to him. I induced the keeper to place him in a small cage to himself, and this he seemed to appreciate very much. When I would go to record his sounds on the phonograph, I held him in one hand, while he would take the tube in his tiny black hands, hold it close up to his mouth, and talk into it just like a good little boy who knew what to do and how to do it. He would sometimes laugh and always chatter to me as long as he could see me. He would sit on my hand and kiss my cheeks, put his mouth up to my ear and chatter just as though he knew what my ears were for. He was quite fond of the head-keeper and also of the director, but he entertained a great dislike for one of the assistant-keepers, and he has very often told me some very bad things about that man, but I could not understand them. I shall long remember how this dear little monk would cuddle up under my chin, and try so hard to make me understand some sad story which seemed to be the burden of his life. He readily understood the sounds of his own speech which I repeated to him, and I have made some of the best records of his voice that I have ever succeeded in making of any monkey, some of which I have preserved up to this time. They present a wide range of sounds, and I have studied them with special care and pleasure because I knew that they were addressed to me in person; and being aware that the little creature was uttering these sounds to me with the hope that I would understand them, I was more anxious to learn just what he really said to me in this record than if it had contained only some casual remark not addressed to me. This little Simian was born in the Amazon Valley in Brazil, and was named for the late Emperor.

[Sidenote: PUCK AND THE PHONOGRAPH]

A short time ago I borrowed from a dealer in Washington a little Capuchin called Puck, and had him sent to my apartments, where I kept a phonograph. I placed the cage in front of the machine upon which I had adjusted the horn, and had placed the record of my little friend Pedro. I concealed myself in an adjoining room, where I could watch the conduct of my subject through a small hole in the door. I had a string attached to the lever of the machine and drawn taut through another hole in the door, so that I could start the machine at any desired moment, and at the same time avoid attracting the attention of the monkey, either by my presence or by allowing him to see anything move. After a time, when everything was quiet, I set the machine in motion and treated him to a phonographic recital by little Pedro. This speech was distinctly delivered through the horn to Puck, from whose actions it was evident that he recognised it as the voice of one of his tribe. He looked at the horn in surprise and made a sound or two, glanced around the room and again uttered a couple of sounds as he retired from the horn, apparently somewhat afraid. Again the horn delivered some exclamations in a pure Capuchin dialect, which Puck seemed to regard as sounds of some importance. He cautiously advanced and made a feeble response, but a quick, sharp sound from the horn seemed to startle him, and failing to find any trace of a monkey, except the sound of a voice, he looked at the horn with evident suspicion, and scarcely ventured to answer any sound it made. When I had delivered to him the contents of the record I entered the room again, and this seemed to afford him some relief.

[Sidenote: PUCK'S VOICE AND ACTIONS]

A little later I adjusted my apparatus for another trial, and this time I hung a small mirror just above the mouth of the horn. Then retiring again from the room I left him to examine his new surroundings, and he soon discovered the new monkey in the glass and began to caress and chatter to it. After a while I started the phonograph again by means of the string, and when the horn began to deliver its Simian oration it appeared to disconcert and perplex Puck. He would look at the image in the glass, then he would look into the horn; he would retire with a feeble grunt and a kind of inquisitive grin, showing his little white teeth, and acting as though in doubt whether to regard the affair as a joke, or to treat it as a grim and scientific fact. His voice and actions were exactly like those of a child, declaring in words that he was not afraid, but betraying fear in every act, and finally blending his feelings into a genuine cry. Puck did not cry, but the evidence of fear made the grin on his face rather ghostly. Again he would approach the mirror, then listen to the sounds which came from the horn, and it appeared from his conduct that there was a conflict somewhere. It was evident that he did not believe that the monkey which he saw in the glass was making the sounds which came from the horn. He repeatedly put his mouth to the glass, and caressed the image which he saw there, and at the same time showed a grave suspicion and some concern about the one which he heard in the horn, and tried to keep away from it as much as possible. His conduct in this case was a source of surprise to me, as the sounds contained in the record which I had repeated to him were all uttered in a mood of anxious, earnest entreaty, which to me seemed to contain no sound of anger, warning, or alarm, but which, on the contrary, I had interpreted as a kind of love speech, full of music and tenderness. I had not learned the exact meaning of any one of the sounds contained in this cylinder, but had ascribed in a collective and general way such a meaning to this speech. But from Puck's conduct I was led to believe that it was a general complaint of some kind against those monkeys in that other cage who had made life a burden to little Pedro. One thing was clear to my mind, and that is that Puck interpreted the actions of the monkey which he saw in the glass to mean one thing, and the sounds which he heard from the horn to mean quite another.

[Sidenote: FORM OF SPEECH USED BY MONKEYS]

I do not think that their language is capable of shaping sentences into narrative or giving any detail in a complaint, for I have never seen anything yet among them which would justify one in ascribing to them so high a type of speech; but in terms of general grievance it may have conveyed to Puck the idea of a monkey in distress, and hence his desire to avoid it; while the image in the glass presented to him a picture of his own mood, and he therefore had no cause to shun it. I do think, however, that the present form of speech used by monkeys is developed far above a mere series of grunts and groans, and that some species among them have a much more copious and expressive form of speech than others. From many experiments with the phonograph, I am prepared to say with certainty that some have much higher phonetic types than others. I have traced some slight inflections which I think beyond a doubt modify the values of their sounds. I find that some monkeys do not make some of these inflections at all, although the phonation of a species is generally uniform in other respects. In some cases it seems to me that the inflections differ slightly in the same species, but long and constant association seems to unify these dialects in some degree, very much the same as like causes blend and unify the dialects of human speech. I have found one instance in which a Capuchin had acquired two sounds which strictly belonged to the tongue of the white-faced Cebus. I was surprised when I heard him utter the sounds, and thought at first that they were common to the speech of both varieties; but on inquiry I found that he had been confined in a cage with the white-face for nearly four years, and hence my belief that he acquired them during that time.

The most remarkable case which has come under my observation is one in which a young white-face has acquired the sound which means food in the Capuchin tongue. This event occurred under my own eyes. I regard this matter as so noteworthy and attended by such conditions as to show that the monkey had a motive in learning the sound, that I shall relate the case in detail.

[Sidenote: THE WHITE-FACED CEBUS]

In the room where the monkeys were kept by a dealer in Washington, there was a cage which contained a young white-faced Cebus of rather more than average intelligence. He was a quiet, sedate, and thoughtful little monk, whose grey hair and beard gave him quite a venerable aspect, and for this reason I called him Darwin. From some cause unknown to me he was afraid of me, and I showed him but little attention. On the same shelf and in an adjacent cage lived the little Capuchin, Puck. The cages were only separated by an open wire partition, through which they could easily see and hear each other. For some weeks I visited Puck almost daily, and in response to his sound for food I always supplied him with some nuts, banana, or other food. I never gave him any of these things to eat unless he would ask me for them in his own speech. On one of my visits my attention was attracted by little Darwin, who was uttering a strange sound which I had never before heard one of his species utter. I did not recognise the sound at first, but very soon discovered that it was intended to imitate the sound of the Capuchin, in response to which I always gave Puck some nice morsel of food. Darwin had undoubtedly observed that this sound made by Puck was always rewarded with something good to eat, and his evident motive was to secure a like reward. After this I always gave him some food in acknowledgment of his efforts, and I observed from day to day that he improved in making this sound, until at last it could scarcely be detected from the sound made by Puck. This was accomplished within a period of less than six weeks from my first visit. In this case, at least, I have seen one step taken by a monkey in learning the tongue of another. This was most interesting to me in view of the fact that I had long believed, and had announced as my belief, that no monkey ever acquired the sounds made by another species, or, indeed, ever tried to do so. I admit, however, that this one instance alone is sufficient to cause me to recede from a conclusion thus rendered untenable, and the short time in which this one feat was accomplished would indicate that the difficulty was not so great as I had regarded it. [Sidenote: SPEECH USUALLY LIMITED] I still regard it as a rule, however, that monkeys do not learn each other's speech, but the rule is not without exceptions. I have observed, and called attention to the fact, that when two monkeys of different species are caged together, that each one will learn to understand the speech of the other, but does not try to speak it as a rule. When he replies at all, it is always in his own vernacular. I wish to impress the fact, that monkeys do not generally carry on a connected conversation. Their speech is usually limited to a single sound or remark, which is replied to in the same manner; and to suppose that their conversations are elaborate or of a highly social character, is to go beyond the bounds of reason. This is the respect in which the masses fail to understand the real nature of the speech of monkeys or other animals.

CHAPTER V.

Five little Brown Cousins: Mickie, Nemo, Dodo, Nigger, and McGinty--Nemo apologises to Dodo.

During the past winter there lived in Central Park a bright, fine, little monkey by the name of Mickie. He did not belong to the Park, but was merely kept as a guest of the city during the absence of his master in Europe. Mickie is a well-built, robust, good-natured monkey of the Capuchin variety. He does not talk much except when he wants food or drink, but he and I are the best of friends, and I frequently go into his cage to have a romp with him and his four little cousins.

When I first began to visit the Park in the fall of 1891, Mickie showed a disposition to cultivate my acquaintance, and as it ripened into a friendship day by day, we found great pleasure in each other's society. As the monkey-house was open to the public at nine o'clock in the morning, I had to make my calls at sunrise or thereabouts, in order to avoid the visitors who daily throng this building.

[Sidenote: NEMO AND MICKIE]

In this cage was kept another little boarder of the same species, which belonged to Mr. G. Hilton Scribner, of Yonkers. The keeper did not know the name or anything of the past history of this little stranger, and for want of some identity and a name I called him Nemo. He was a timid, taciturn little fellow, quite intelligent, and possessed of an amount of diplomacy equal to that of some human beings. He was the smallest monkey in the cage, on which account he was somewhat shy of the others. He was thoughtful, peaceable, but full of "guile." He sought on all occasions to keep on the best terms with Mickie, to whom he would toady like a sycophant. He would put his little arms about Mickie's neck and hang on to him in the most affectionate manner. He would follow him like a shadow, and stay by him like a last hope. If anything ever aroused the temper of Mickie it was sure to make Nemo mad too; if Mickie was diverted and would laugh, Nemo would laugh also if he was suffering with a toothache. He was as completely under the control of Mickie as the curl in Mickie's tail. When I first began to visit them Nemo would see Mickie bite my fingers while we were playing, and he supposed it was done in anger. Nemo never lost a chance to bite my fingers, which he would always do with all his might, but his little teeth were not strong enough to hurt me very much. He would only do this after seeing Mickie bite me, and he did not evince any anger in the act, but appeared to do so merely as a duty. He would sneak up to my hands and bite me unawares; then he would run to Mickie and put his arm about his neck just as you have seen some boys do when trying to curry favour with a larger boy. On one occasion while in the cage with them he slipped up to me and bit my finger, for which I kindly boxed his little ears. I would then give Mickie my finger and allow him to bite it, after doing which I slapped him gently and then give it to him again. I would then allow Nemo to bite my finger, and if he bit it too hard I would slap him again, and in this manner soon taught him to understand that Mickie only bit me in fun, and he evidently learned that this was a fact. He did not appear, however, to catch the point clearly or see any reason therefor, but on all occasions thereafter he would take my finger in his mouth and hold it in his teeth, which were scarcely closed upon it. This he would do for a minute at a time without having the least apparent motive except that he had seen Mickie do so. [Sidenote: MICKIE'S ATTACHMENT] Often while holding my finger in this manner, with a look of seriousness worthy of a supreme judge, he would roll his little eyes at me in the most inquiring manner, as if to say "how is that"? When he once realised that Mickie was so much attached to me, Nemo always showed a desire to be on friendly terms with me; and when I would go into the cage to play with Mickie and McGinty, he always wanted to be counted in the game. When I had anything for them to eat he always wanted a seat of honour at the table, and he would at times want to fight for me when the other monkeys got too friendly. Poor little fellow, he is now dead, but the image of his cute little face and original character are deeply imprinted on my mind. I was never able to secure a record of the sounds of his little voice, though I have often heard him talk. He had a soft musical voice, and great power of facial expression.

[Sidenote: APOLOGY TO DODO]

One of the most remarkable things I have ever observed among monkeys was done by this little fellow. On two separate occasions I have seen him apologise to Dodo in the most humble manner for something he had done, and I tried very hard to secure a record of this particular speech, in which I totally failed, as I could not foreknow when such an act would be done, and therefore could not have my phonograph in place to obtain such a record. I called the attention of Mr. F. S. Church, the eminent artist, to this act, with the hope that he might be able to make a sketch of Nemo while in this attitude. I do not know what the offence was, but the pose and expression as well as the speech were very impressive. He sat in a crouching position, with the left hand clasping the right wrist, and delivered his speech in a most energetic but humble manner. The expression on his face could not be misunderstood. After a few moments he paused briefly, and then seemed to repeat the same thing some two or three times. The manner of his delivery was very suggestive, and his demeanour was conciliatory. When he had quite finished his speech, Dodo, to whom the apology was being made, and who had listened to it in perfect silence, delivered a sound blow with her right hand on the left side of the face of the little penitent, to which he responded with a soft cry, while Dodo turned and left him without further debate. I also called the attention of the keeper to this act, and he assured me that he had repeatedly witnessed the same. What the subject of his speech was or the cause which brought it about I am not able to say, nor can I say with certainty to what extent he explained, but that it was an apology, or explanation of some kind at least, I have not the slightest doubt. I do not believe, of course, that his speech contained any details concerning the offence, but that it expressed regret, penitence, or submission does not to my mind admit of a doubt. I have seen a few other cases somewhat similar to this, but none of them comparing in point of polish and pathos to that of Nemo in his unique little speech.

Nigger was of this same species: he was in poor health most of the winter, being afflicted with some spinal trouble. But, notwithstanding his affliction, he was a good talker. His infirmity, however, placed him at the mercy of the other inmates of the cage, and as monkeys are naturally cruel and entirely destitute of sympathy, the daily life of Nigger could not be expected to be a very happy one. From this state of facts Nigger usually kept to himself, and was not intimate with any other monkey in the cage. I have frequently given Nigger some choice bits of food while I was in the cage, and protected him from the other monkeys while he was eating it. This he seemed to fully appreciate, and always located himself at a certain point in the cage where his defence could be effected with the least difficulty. Nigger frequently indulged in the most pathetic and touching appeals to his keeper, and went through many of the gestures, sounds, and contortions which will be described in the next chapter, as a part of the speech and conduct of Dodo, some of whose remarkable poses and expressions have been faithfully portrayed by Mr. Church.

[Sidenote: McGINTY AT CENTRAL PARK]

Among my personal friends of the Simian race, there is none more devoted to me than little McGinty, another winter boarder at Central Park. From the first of my acquaintance with McGinty we had been staunch friends, and when I go to visit him he expresses the most unbounded delight. He will reach his little arms through the bars of the cage, and put his hands on my cheeks, hold his mouth up to the wires, and talk to me at great length. When I go into the cage he will place himself on a perch in the cage, where he will sit with his arms around my neck, lick my cheeks affectionately, pull my ears, and chatter to me in a sweet but plaintive tone. When Mickie joins the play, which he invariably does, by climbing or jumping on to my shoulders, and interrupting the t 㑥 e-?t 㑥 e between McGinty and myself, poor little McGinty's jealousy, which is his supreme passion, causes him to retire in disgust, and he will sometimes pout for several minutes without even accepting food from me. After he has pouted for a while, however, he will sometimes make overtures of reconciliation and seek by various means to divert my attention. One of his favourite means of renewing favour with me, was to whip poor little Nigger. He would look at me and laugh, grin and make grimaces, and then dash off at Nigger and want to eat him up. He did not seem to understand why I objected to this whipping Nigger. Monkeys do not regard it as a breach of honour to whip the helpless and feeble members of their tribe. They are not unlike a large percentage of mankind. They always hunt for easy prey, and want to fight something that is easily whipped. They are not great cowards, but when once whipped they rarely attempt the second time to contest matters with their victors. [Sidenote: CAGE OF CAPUCHIN MONKEYS] In this cage, containing five brown Capuchin monkeys, it was not difficult to see that Mickie ran things to suit himself. McGinty was the only one of the four in the cage with him that ever contested any right with Mickie, and for a long time it was a question in my mind who was to win in the end. The next to them in authority was Dodo, who never attempted to control Mickie or McGinty, but always made Nemo and Nigger stand about. Fourth in line of authority was Nemo, who always resented any offence from others by making Nigger take a corner; and the only victims that Nigger had were the little white-faces, which never fight anything and are always on the run. When it was finally decided between Mickie and McGinty that Mickie should be captain, McGinty readily accepted the place of first lieutenant, which rank he has continued to hold without challenge. When once the question is settled among the cage of Simians, the debate does not appear to be renewed at any future time. They never go to court with their grievances, and rarely appeal a second time to force when the question has once been decided against them. Some human beings might profit by studying this trait of monkeys.

CHAPTER VI.

Dago Talks about the Weather--Tells me of his Troubles--Dodo in the "Balcony Scene"--Her Portrait by a great Artist.

On one of my visits to Chicago, in the autumn of 1890, I went to pay my respects to Dago, the little brown monkey in Lincoln Park. He had been sick for a while, and had not fully recovered, although he was able to receive visitors, and his appetite for pea-nuts was fairly well restored. On the morning of which I speak, it was dark and stormy. A fierce wind and terrible rain prevailed from the north-west. I went to the building just after daylight, in order to be alone with the monkey, and when I entered the house, Frenchie, the head-keeper, told me how very sick little Dago had been since I had left him on the day before. I approached the cage and began to caress him, to which he replied in low whimpering tones, as though he understood the nature of what I was saying to him. Presently he raised himself erect upon his hind feet, and placing his hands on his side, pressed and rubbed it as though he was in great pain, and uttered some sounds in a low, piping voice.

The sound itself was pathetic, and when accented by his gestures, it was really very touching. [Sidenote: DAGO AND THE WEATHER] At this juncture, a hard gust of wind and rain dashed against the window near his cage, whereupon the little monk turned away from me, ran to the window and looked out, and uttered a sound quite different from the ones he had just been delivering to me. Still standing erect, he appeared deeply interested, and stood for a few moments at the window, during which time he would turn his head towards me and utter this sound. That the sound he uttered was addressed to me could not be doubted, and his manner in doing so was very human-like. Then returning to me, still standing erect, he would renew this plaintive speech in the most earnest manner, and continue it until another gust would call him to the window. I observed that each time he went to the window he uttered the same sound, as well as I could detect by ear, and would stand for some time watching out of the window, and occasionally turn his head and repeat this sound to me. When returning to me again, he would resume his sad story, whatever it was. I secured a good record of that part of his speech which was made when near me at the front of the cage, but the remarks made while at the window were not so well recorded, yet they were audible, and I reproduced them on the phonograph at a subsequent visit. My opinion was that the sound he uttered while at the window must allude in some way to the state of the weather, and this opinion was confirmed by the fact that on a later occasion, when I repeated the record to him, the weather was fair; but when the machine repeated those sounds which he had uttered at the window on the day of the storm, it would cause him to turn away and look out of the window; while at the other part of the record he evinced but little interest, and, in fact, seemed rather to avoid the phonograph as though the sounds suggested something which he disliked. I am quite sure that the remarks which he made to me at the front of the cage were a complaint of some kind, and, from its intonation and the manner in which it was delivered, I believed that it was an expression of pain. It occurred to me that the state of the weather might have something to do with his feelings, and that he was conscious of this fact, and desired to inform me of it.

About a year from that time, I became quite intimate with a feeble little monkey, which is described elsewhere by the name of Pedro, and of whose speech I made a good record. The sounds of his speech so closely resembled those made by Dago, that I was not able to see that they differed in any respect, except in loudness. Unfortunately, the cylinders containing Dago's record had been broken in shipping, and I was therefore unable to compare the two by analysis; but the sounds themselves resembled in a striking degree, and the manner of delivery was not wholly unlike, except that Pedro did not assume the same pose nor emphasise them with the same gestures.

[Sidenote: DODO, THE JULIET OF THE TRIBE]

During my stay in New York the past winter, I have been frequently entertained by a like speech from little Dodo, who was the Juliet of the Simian tribe. She belonged to the same species as the others, but her oratory was of a type far superior to that of any other of its kind that I have ever heard. At almost any hour of the day, at the approach of her keeper, she would stand upright and deliver to him the most touching and impassioned address. The sounds which she used, and the gestures with which she accented them, as far as I could determine, were the same as those used by Dago and Pedro in their remarks to me as above described, except that Dodo delivered her lines in a much more impressive manner than either of the others. [Sidenote: DODO AND HER KEEPER] I asked the keeper to go into the cage with me, and see if he could take her into his hands. We entered the cage, and after a little coaxing she allowed him to take her into his arms, and after caressing her for a while, and assuring her that no harm was meant, she would put her slender little arms about his neck, and cuddle her head up under his chin like an injured child. She would caress him by licking his cheeks and chattering to him in a voice full of sympathy, and an air of affection worthy of a human being. During most of this time she would continue her pathetic speech without a moment's pause, and was not willing under any conditions to be separated from him. The only time at which she would ever show any anger at me, or threaten me with assault, would be when I would attempt to lay hands on her keeper, or release him from her warm embrace. At such times, however, she would fly at me with great fury, and attempt to tear my very clothes off, and on these occasions she would not allow any other inmate of the cage to approach him, or to receive his attention or caresses. The sounds which she uttered were pitiful at times, and the tale she told must have been full of the deepest woe. I have not been able up to this time to translate these sounds literally, but their import cannot be misunderstood. My belief is that her speech was a complaint against the inmates of the cage, and that she was begging her keeper not to leave her alone in that great iron prison, with all those big, bad monkeys, who were so cruel to her. One reason for believing this to be the nature of her speech, is that in all cases where I have heard this speech and seen these gestures made, the conditions were such as to indicate that such was its nature. It has, however, every appearance of love-making of the most intense type. It is quite impossible to describe fully and accurately the sounds, and much more so the gestures, made on these occasions, so that the reader would be impressed as with the real act and speech. Dodo would stand erect on her feet, cross her hands on her heart, and in the most touching but graceful manner go through with the most indescribable contortions; she would sway her body from side to side, turn her head in the most coquettish manner, and move her folded hands dramatically, while her face would be adorned with a Simian grin of the first order, and the soft, rich notes of her voice were perfectly musical. She would bend her body into every graceful curve that can be imagined, move her feet with the grace of the minuet, and continue her fervent speech as long as the object of her admiration appeared to be touched by her appeals. Her voice would range from pitch to pitch and from key to key, and, with her arms folded, she would glide across the floor of her cage with the grace of a ballet girl; and I have seen her stand with her eyes fixed upon her keeper, and hold her face in such a position as not to lose sight of him for a moment, and at the same time turn her body entirely around, in her tracks, with the skill which no contortionist has ever attained. [Sidenote: MONKEYS SHED TEARS] During these orations I have observed the little tears standing in the corner of her eyes, which indicated that she herself must have felt what her speech was intended to convey. These little creatures do not shed tears in such abundance as human beings do, but they are real tears, and are doubtless the result of the same causes that move the human eyes to tears.

It has been my experience that these sounds appeal directly to our better feelings. What there is in the sound itself I cannot say, but it touches some chord in the human heart which vibrates in response to it. It has impressed me with the thought that all our senses are like the strings of some great harp, each one having a certain tension; so that any sound produced through an emotion would find response in that chord which is in unison with it. Indeed, I have thought that our emotions and sensations may be like the diatonic scale in music, and that the organs through which they act may respond in tones and semitones, and that each multiple of any fundamental tone will affect the chord in unison with it, like the strings upon a musical instrument. The logical deduction thence would be, that our sympathies and affections are the chords, and our aversions and contempt the discords, of that great harp of passion.

CHAPTER VII.

Interpretation of Words--Specific Words and Signs--The Negative Sign and Sounds--Affirmative Expressions--Possible Origin of Negative and Positive Signs.

In my intercourse with these little creatures, I cannot forget how often I have caught the spirit of their tones when no ray of meaning as mere words of speech had dawned upon me, and it is partly through such means that I have been able to interpret them. As a rule, each act of a monkey is attended by some sound, and each sound by some act, which, to another monkey of the same species, always means a certain thing. There are many cases, perhaps, in which acquired words or shades of dialect are not quite clear to them, just as we often find in human speech; but monkeys appear to meet this difficulty and overcome it, just as men do. They talk with one another on a limited number of subjects, but in very few words, which they frequently repeat if necessary. Their language is purely one of sounds, and while those sounds are accompanied by signs, as a rule, I think they are quite able to get along better with the sounds alone than with the signs alone. The rules by which we may interpret the sounds of Simian speech are the same as those by which we would interpret human speech. If you should be cast away upon an island inhabited by some strange race of people whose speech was so unlike your own that you could not understand a single word of it, you would watch the actions of those people and see what act they did in connection with any sound they made, and in this way you would gradually learn to associate a certain sound with a certain act, until at last you would be able to understand the sound without seeing the act at all; and such is the simple line I have pursued in the study of the speech of this little race--only I have been compelled to resort to some very novel means of doing my part of the talking. Since I have been so long associated with them, I have learned to know in many cases what act they will perform in response to certain sounds; and as I grow more and more familiar with these sounds, I become better able to distinguish them, just as we do with human speech.

Until recently, I have believed that their sounds were so limited in number as to preclude any specific terms in their vocabulary; but now I am inclined to modify this opinion somewhat, as I have reason to believe that they have some specific terms--such as a word for monkey, another word for fruit, and so on. They do not specify, perhaps, the various kinds of monkeys; but monkeys in general, in contradistinction to birds or dogs. Their word for fruit does not specify the kind, but only means fruit in a collective sense, and only as a kind of food. I am not positive as yet that their specific terms may even go so far as this, but I infer that such may be the case from one fact which I have observed in my experience. When I show a monkey his image in a mirror, he utters a sound on seeing it, especially if he has been kept away from other monkeys for a long time; and all monkeys of the same species, so far as I have observed, under like conditions use the same sound and address it in the same way to the image in the glass. In a few instances I have seen strange monkeys brought in contact with each other, and have observed that they use this same sound on their first meeting. The sound is always uttered in a low, soft tone, and appears to have the value of a salutation. When kept in a cage with other monkeys, they do not appear to salute the image in the glass, but chatter to it, and show less surprise at seeing it than in cases where they have been kept alone for some time.

In cases where monkeys have been fed for a long time on bread and milk, or on any one kind of food, when a banana is shown him he uses a sound which the phonograph shows to differ slightly from the ordinary food sound. I have recently had reason to suspect that this difference of inflection somewhat qualifies the sound, and has a tendency to make it more specific. The rapidity with which these creatures utter their speech is so great that only such ears as theirs can detect these very slight inflections. I am now directing my observations and experiments to this end, with the hope that I may determine with certainty in what degree they qualify their sounds, by inflections or otherwise. I have observed that in the phonograph the sounds which formerly appeared to me to be the same are easily distinguished when treated in the manner described in the second part of this work, where I describe at length some of my experiments with this wonderful machine.

One of the most certain of my discoveries in the Simian speech, is the negative sign and the word "no." The sign is made by shaking the head from side to side in a fashion almost exactly like that used by man to express the same idea. I have no longer any doubt of the intent and meaning of this sign, and the many tests to which I have subjected it compel me accept the result as final.

A little more than a year ago, my attention was called to this sign by the children who own the little Capuchin, Jack, in Charlestown. A number of times they said to him in my presence, "Jack, you must go to bed." At which he would shake his little black head, as if he really did not wish to comply. I watched this with great interest; but it was my belief at that time that he had been trained to do this, and that the sign did not really signify to him anything at all. The children, however, declared to me that he really meant "no." To believe that he meant this would presuppose that he understood the combination of words quoted; and this was beyond the limits of my faith, although it was certain that a repetition of the sentence always elicited from him the same sign, which indicated that he recognised it as the same sentence or combination of sounds, and gave it the same reply each time. I concluded that he had been taught to associate this sign with some sound-- for instance, "bed" or "go"; but since that time I have found the sign to be almost universal with this species of monkey, and they use the sign to express negation. I have seen them use the sign in response to certain things which were wholly new to them, but where the idea was clear to them and they desired to express dissent. The fact that this sign is common to both man and Simian, I regard as more than a mere coincidence; and I believe that in this sign I have found the psycho-physical basis of expression.

I have made scores of experiments on this subject, and I find this sign a fixed factor of expression. In one case, where I tried to induce a monkey to allow me to take him into my hands from the hand of his master, he would shake his head each time, and make a peculiar sound somewhat like a suppressed cluck. I would try to coax him with nuts, in response to which he would make the same sound and sign each time, and his actions showed beyond all controversy his intention. I had taught a monkey to drink milk from a bottle by sucking it through a rubber nipple, and after he had satisfied his thirst, when I would try to force the bottle to his lips, he would invariably respond by a shake of the head in the manner described, and at the same time utter a clucking sound. I tried many similar experiments with three or four other monkeys, and secured the same result in each case. In another instance, where a monkey was confined in a small cage so that I could easily catch him in order to tame him by handling, when I would put my hand into the cage to catch him, he would shake his head in this manner and accompany the act by a plaintive sound which was so touching, that I could not obtain my own consent to persecute the little prisoner by compelling him to submit to my caresses. I have found that the little rogue, McGinty, in Central Park does the same thing at times when I go into the cage and attempt to put my hands on him, and especially when he has taken refuge in a corner to nurse his jealousy. While I remain outside the cage, he is so devoted to me that he will scarcely leave me to get something to eat; but when I enter the cage, and reach out my hand toward him, he will shake his little head and utter that peculiar clucking sound. Many of these tests I have repeated over and over with the same results, and, noting the conditions at the time, I am thoroughly convinced that the sign and sound mean "no." I have observed that this sign is always made in the same manner; but sometimes it is accompanied by a clucking sound, while at other times it is a soft whimpering sound, almost like a low plaintive whistle. [Sidenote: SIGN USED WITHOUT SOUND] The sign is frequently used without the sound at all, and I must impress it upon my reader that these results do not always present themselves in every experiment, as much depends upon the mood and surroundings of the subject. I have found that one advantage is to have the monkey confined in a very small cage, as otherwise he will turn away and get out of your reach when you press anything upon him that he does not want. I have also found much better results by having the monkey alone, and where he can neither see nor hear other monkeys.

Having discovered the sign of negation among the Simians, I began an investigation to ascertain how far it could be found among the races of mankind. I have carried my search far beyond the limits of local inquiry, and up to this time I have found only a few trifling exceptions in the use of this sign among all the races of men, and those few exceptions are found among the Caucasian race, and appear to be confined to Southern Europe. I have heard that among certain island tribes of Polynesia these signs are reversed, but I have been assured by two officers of the English navy and two of the United States navy, who have visited the islands in question, that such is not the case. Among the Indians, Mongolians, and Negroes I have found no noteworthy exceptions. I have inquired among mothers who have raised families to ascertain when they first observed this sign as an expression among their children; and from the consensus of opinion it appears that this is about the first sign used by infants to express negation.

[Sidenote: THE POSITIVE SIGN]

I have not found the positive sign, or sign of affirmation, by a nod of the head, to be so general, yet it has a wide range within the human family, and appears to be used to some extent among the lower primates.

Seeking a source from which these signs may have originated, I have concluded that they may arise from two circumstances. The negative sign doubtless comes from an effort to turn the head away from something which is not desired, and that with such an intent it has gradually crystallised into an instinctive expression of negation or refusal; while the nod of affirmation or approval may have grown out of the intuitive lowering of the head, as an act of submission or acquiescence, or from reaching the head forward to receive something desired, or they may have come from these two causes conjointly.

[Sidenote: ALPHABET FOR SIMIAN SPEECH]

This is only one of a great many points in which the speech of Simians coincides with that of man. It is true we have no letters in our alphabet with which to represent the sounds of their speech, nor have we the phonetic equivalence of their speech in our language; but it is also true that our alphabet does not fully represent or correctly express the entire phonetic range of our own speech; but the fact that our speech is not founded upon the same phonetic basis, or built up into the same phonetic structures, is no reason that their speech is not as truly speech as our own. That there are no letters in any alphabet which represent the phonetic elements of Simian speech, is doubtless due to the fact that there has never been any demand for such; but the same genius which invented an alphabet for human speech, actuated by the same motives and led by the same incentives, could as easily invent an alphabet for Simian speech. It is not only true that the phonetic elements of our language are not represented by the characters of our alphabet, but the same is true to some extent of our words, which do not quite keep pace with human thought. In the higher types of human speech there are thousands of words and ideas which cannot be translated into or expressed by any savage tongue, because no savage ever had use for them, and no savage tongue contains their equivalence. The growth of speech is always measured by the growth of mind. They are not always of the same extent, but always bear a common ratio. It is a mental product, and must be equal to the task of coining thoughts into words. It is essential to all social order, and no community could long survive as such without it. It is as much the product of mind and matter as salt is the product of chlorine and sodium.

CHAPTER VIII.

Meeting with Nellie--Nellie was my Guest--Her Speech and Manners--The little Blind Girl--One of Nellie's Friends--Her Sight and Hearing--Her Toys, and how she Played with them.

One of the most intelligent of all the brown Capuchins that I have ever seen was Nellie, who belonged to a dealer in Washington. When she arrived there, I was invited to call and see her. I introduced myself in my usual way, by giving her the sound for food, to which she promptly replied. She was rather informal, and we were soon engaged in a chat on that subject, the one above all others that would interest a monkey. On my second visit she was like an old acquaintance, and we had a fine time. On my third visit she allowed me to put my hands into her cage, and handle her with impunity. On my next visit I took her out of the cage, and we had a real romp. This continued for some days, during which time she would answer me on all occasions when I used the word for food or drink. She had grown quite fond of me, and always recognised me as I entered the door. [Sidenote: NELLIE AND THE BLIND GIRL] About this time there came to Washington a little girl who was deaf, dumb, and blind; she was accompanied by her teacher, who acted as her interpreter. One of the greatest desires of this little girl's life was to see a live monkey-- that is, to see it with her fingers. The dealer who owned the monkey sent for me to come down and show it to her, as I could handle the monkey for her. I took Nellie from the cage, and when any one except myself would put hands upon her she would growl and scold and show her temper; and when the little blind girl first attempted to put her hands on her, Nellie did not like it at all. I stroked the child's hair and cheeks with my own hand first, and then with Nellie's; she looked up at me in an inquiring manner, and uttered one of those soft, flute-like sounds a few times, and then began to pull at the cheeks and ears of the child. Within a few moments they were like old friends and playmates, and for nearly an hour they afforded each other great pleasure, at the end of which time they separated with reluctance. The little Simian acted as if she was conscious of the sad affliction of the child, but seemed at perfect ease with her, although she would decline the tenderest approach of others. She would look at the child's eyes, which were not disfigured, but lacked expression, and then look up at me as if to indicate that she was aware that the child was blind, and the little girl appeared not to be aware that monkeys could bite at all. It was a beautiful and touching scene, and one in which the lamp of instinct shed its feeble light on all around.

On the following day, by an accident in which I really had no part, except that of being present, Nellie escaped from her cage, and climbed up on a shelf occupied by some bird-cages. As she attempted to climb up, of course the light wicker cages with their little yellow occupants fell to the floor by the dozen. I tried to induce her to return or to come to me, but the falling cages, the cry of the birds, the talking of parrots, and the scream of other monkeys, frightened poor Nellie almost out of her wits. Thinking that I was the cause of her trouble, because I was present, she would scream with fright at my approach. She was not an exception to that general rule which governs monkeydom, which is to suspect every one of doing wrong except itself.

I had her removed to my apartment, where I supplied her with bells and toys, and fed her on the fat of the land; and by this means we slowly knitted together the broken bones of our friendship once more. But when once a monkey has grown suspicious of you they never recover entirely from it, it seems, for in every act thereafter, however slight, you can readily see that they suspect you of it; but with great care and caution you can make them almost forget the trouble. While I kept Nellie at my rooms I made some good records of her speech on the phonograph, and studied her with special care; but as the province of this work is the speech of that little race, I must forego the pleasure of telling some intensely funny things with which she entertained me, excepting so far as they are relevant to speech.

[Sidenote: NELLIE'S FONDNESS FOR A LITTLE BOY]

A frequent and welcome visitor to my study was a bright little boy, about six years old, for whom Nellie entertained a great fondness, as she also did for my wife. At the sight of the boy Nellie would go into perfect raptures, and when he would leave her, she would call him so earnestly and whine so pitifully that one could not refrain from sympathy. On his return she would laugh audibly, and give every sign of extreme joy. She never tired of his company, nor gave any part of her attention to others when he was present. Some children living next door always found great delight in calling to see Nellie, and she always showed her pleasure at their visits. On these occasions, Nellie made it a point to entertain them, and always showed herself to the best advantage. When I wished to make a good record of her sounds, and especially of her laughter, I always brought the little boy to my aid. The boy would conceal himself in the room, and after Nellie had called him a few times he would jump out from his place of concealment and surprise her, whereupon she would laugh till she could be heard through the whole house; and in this manner I secured some of the best records I have ever made of the laughter of any monkey. When the boy would conceal himself again, I secured the peculiar sound with which she would try to attract his attention. The sound which she used in calling him or my wife was unlike that which she made for any other purpose; and while it is difficult to say whether the grammatical value of this sound is that of a noun or of a verb, it is evident that it was used for the special purpose of calling or attracting attention. If its value is that of a noun, it has not, in my opinion, any specific character, but a term which would be applied alike to boys, monkeys, horses, birds, or any other thing which she might desire to call. If in its nature it is a verb, it is equivalent to the name of the act, and combines the force of the imperative and infinitive moods.

[Sidenote: EMOTIONS OF MAN AND SIMIAN]

The uniform expression of the emotions of man and Simian is such as to suggest that, if thought was developed from emotion and speech was developed from thought, that the expressions of emotion were the rudiments from which speech is developed.

A striking point of resemblance between human speech and that of the

Simian is found in a word which Nellie used to warn me of approaching danger. It is not that sound which I have elsewhere described as the alarm-sound, and which is used only in case of imminent and awful danger; but this sound is used in case of remote danger or in announcing something unusual. As nearly as I can represent the sound by letters, it would be "e-c-g-k," and with this word I have been warned by these little friends many times since I first heard it from Nellie.

[Sidenote: NELLIE'S ACTIONS ALMOST HUMAN]

In the following experiment this sound was used with great effect. Nellie's cage occupied a place in my study near my desk. She would stay awake at night as long as the light was kept burning, and as I have always kept late hours, I did not violate the rule of my life in order to give her a good night's rest. About two o'clock one morning, when I was about to retire, I found Nellie wide awake. I drew my chair up to her cage, and sat watching her pranks as she tried to entertain me with bells and toys. I tied a long thread to a glove, which I placed in a corner of the room at a distance of several feet from me, but without letting her see it. I held one end of the string in my hand, I drew the glove obliquely across the floor towards the cage. When I first tightened the string, which I had drawn across one knee and under the other, the glove moved very slightly, and this her quick eye caught at the very first motion. Standing almost on tip-toe, her mouth half open, she would peep cautiously at the glove, and then in a low whisper would say "e-c-g-k"! And every second or so would repeat it, at the same time watching me, to see whether I was aware of the approach of this goblin. Her actions were almost human, while her movements were as stealthy as those of a cat. As the glove came closer and closer she became more and more demonstrative, and when at last she saw the monster climbing up the leg of my trousers, she uttered the sound aloud and very rapidly, and tried to get to the object, which she evidently thought was some living thing. She detected the thread with which I drew the glove across the floor, but seemed in doubt as to what part it played in this act. I saw her eyes several times follow the thread from my knee to the glove, but I do not think she discovered what caused the glove to move. Having done this for a few times, however, with about the same result each time, I relieved her anxiety and fright by allowing her to examine the glove, which she did with marked interest for a moment and then turned away. I tried the same thing over again, but failed to elicit from her the slightest interest after she had examined the glove.

[Sidenote: SOUND OF WARNING]

It will be observed that when Nellie first discovered the glove moving on the floor, as she attempted to call my attention in a low whisper, and as the object approached me she became more earnest, and uttered the sound somewhat louder, and when she discovered the monster, as she regarded it, climbing up my leg, she uttered her warning in a loud voice, not a scream or a yell, but in a tone sufficiently loud for the distance over which the warning was conveyed. The fact of her whispering indicates that her idea of sound was well defined; her purpose was to warn me of the approaching danger without alarming the object against which her warning was intended to prepare me; and as the danger approached me, her warning became more urgent, and when she saw the danger was at hand her warning was no longer concealed or restrained.

Another sound which these little creatures use in a somewhat similar manner, is a word which may be represented by the letters "c-h-i." The "c-h" is guttural like the final "ch" in German, and "i" short like the sound of "i" in hit. This sound is used to give warning of the approach of something which the monkey does not fear, such as approaching footsteps or the sound of voices; and this sound Nellie always used to warn my wife of my approach when I was coming up the stairway. The rooms which I occupied while I kept Nellie were located on the second floor, and the dining-room was on the ground-floor; and hence there were two flights of stairs between, both of which were carpeted. So acute was her sense of hearing, that she would detect my footsteps on the lower stairway, and warn my wife of my approach. She manifested no interest, as a rule, in the sounds made by other persons passing up and down the stairway, which indicated that she not only heard the sounds of my footsteps but recognised them. The first intimation she would give of my coming was always in a whisper. She would first make the sound "c-h-i," and then she would stop and listen. She would repeat the sound and listen again, and as I would approach the door in the hall she would lift her voice to its natural pitch, and utter this sound three or four times in quick succession; and when I turned the door-knob she would show some excitement, and when I entered the room she would always express her satisfaction with a little chuckle. This sound she did not use except to announce something of which she was not afraid, but when she apprehended danger from the cause of the sound, she would use the word "e-c-g-k," and when greatly alarmed she would use the sound which I have described in the former chapter as that of intense alarm or assault.

[Sidenote: MONKEYS DO NOT TALK WHEN ALONE]

Nellie was an affectionate little creature, and could not bear to be left alone, even when supplied with toys and everything she wanted to eat. When she would see me put on my overcoat, or get my hat and cane, she knew what it meant; and when she would see my wife, to whom she was much devoted, put on her cloak and bonnet, she at once foresaw that she would be left alone. Then she would plead and beg and chatter, until she sometimes dissuaded my wife, and she seemed aware that she had accomplished her purpose. I have watched her by the hour, through a small hole in the door, and when quite alone she would play with her toys in perfect silence, and sometimes for hours together she would not utter a single word. She was not an exception to the rule which I have mentioned heretofore, that monkeys do not talk when alone, or when it is not necessary to their comfort or pleasure; and while I am aware that their speech is far inferior to human speech, yet in it there is an eloquence that soothes, and a meaning that appeals to the human heart.

CHAPTER IX.

Affections--A little Flirtation--Some of my personal Friends.

Nellie had spent much of her life in captivity and had been used to the society of children, for whom she showed the greatest fondness, and rarely ever betrayed the slightest aversion to any of them. She delighted to pat their cheeks, pull their ears, and tangle their hair. One of her favourite pastimes was to pull the hairpins out of my wife's hair so that she could get hold of it the better to play with, and my wife has often remarked that Nellie would make an excellent lady's-maid. She would clean one's finger-nails with the skill of a manicure. She would pick every shred, ravelling, or speck from one's clothing. Her aversions and attachments were equally strong. She was not selfish in selecting her friends, nor did she seem to be influenced by age or beauty.

[Sidenote: MONKEYS SHOULD HAVE TOYS]

To let her out of her cage and give her something to play with was happiness enough for her, and I almost think she preferred such a life to the freedom of her Amazon forests. But you cannot afford to turn one out of the cage in a room where there is anything that can be torn or broken, as they enjoy such mischief in the highest degree. Nellie would beg me so piteously to be taken from her little iron prison that I could not have the cruelty to refuse her, even at the cost of some trouble in preparing the room for her; and as we retain these little captives against their will, and treat them worse than slaves by keeping them in close confinement, I think we should at least try to amuse them. It is true they do not have to toil, but I think it would be more humane to make them work in the open air than to confine them so closely, and then deprive them of every source of pleasure. As an act of humanity and simple justice, I would impress upon those who keep such little pets how important a thing it is to keep them supplied with toys. They are just like children in this respect, and for a trifle one can furnish them with all the toys they need. It is cruel, absolutely cruel, to keep these little creatures confined in solitude and deny them the simple pleasure they find in playing with a bell, ball or marbles; and besides this, a trifling outlay in this way will very much prolong their lives. A monkey is always happy if he has something to play with and plenty to eat. [Sidenote: NELLIE WITH THE MATCH-BOX] I do not know of any investment of mine which ever yielded such a great return in pleasure as one little pocket match-safe which cost me twenty-five cents, and which I gave to Nellie one evening to play with. I had put into it a small key to make it rattle, and also some bits of candy. She rattled the box, and found some pleasure in the noise it made. I showed her a few times how to press the spring in order to open it, but her little black fingers were not strong enough to release the spring and make the lid fly open. However, she caught the idea, and knew that the spring was the secret which held it; and when she found that she could not open it with her fingers, she tried it with her teeth. Failing in this, she turned to the wall, and standing upright on the top of her cage, she took the box in both hands and struck the spring against the wall until the lid flew open. She was perfectly delighted at the result, and for the hundredth time at least I closed the box for her to open again. On the following day, when some friends came in to visit her, I gave her the match-box to open again. On this occasion, however, she was in her cage and could not reach the wall through its meshes, and hence had nothing against which to strike the spring to force it open. After looking around her in all directions and striking the box against the wires of her cage a few times, she discovered a block of wood in her cage about six inches square by an inch thick, and this she took and mounted her perch. Balancing the block on the perch she held it with the left foot, while with her right foot she held on to the perch, and with her tail wound through the meshes of her cage to steady herself, she carefully adjusted the match-box in her hands in such a manner as to protect her fingers from the blow. Then striking the spring against the block of wood the lid flew open, and she fairly screamed with delight, and held the box up with pride, wanting me to close the lid again, in order that she might open it.

Finding that the late hours which I kept were beginning to tell on Nellie, and that during the day from time to time I would catch her taking a little nap, I concluded to use some curtains around her cage to avoid disturbing her rest. I drew them around the cage, lapped them over, and pinned them down in front. Then I turned down the light and kept quiet for a while to allow her to go to sleep. After the lapse of a few minutes, I slowly turned up the light and resumed my writing. In an instant I heard the curtains rustle, and looked around, and there I saw her little brown eyes peeping through the folds of the curtains, which she held apart with her little black hands. When she saw what it was that caused all this disturbance, she chattered to me in her soft rich tones, and tried so hard to pull the curtains apart that I removed them from her cage so that she could look around the room. To see her holding the curtains apart in that graceful manner, turning her head from side to side, peeping and smiling at me, and talking in such low tones, was so much like a real flirtation that one who has not seen the like cannot fully appreciate it. And only those who have experienced the warm and unselfish friendship of these little creatures can realise how strong the attachment becomes. When once you enjoy the confidence of a monkey, nothing can shake it, except some act of your own, or one at least which they attribute to you. Their little ears are proof against gossip, and their tongues are free from it.

[Sidenote: THE LOVE OF MONKEYS]

Among the little captives of the Simian race who spend their lives in iron prisons to gratify the cruelty of man, and not to expiate some crime committed or inherent, I have many little friends to whom I am attached, and whose devotion to me is as warm and sincere, so far as I can see, as that of any human being. I must confess that I cannot discern in what intrinsic way the love they have for me differs from my own for them. I cannot see in what respect their love is less divine than is my own. I cannot see in what respect the affections of a dog for a kind master differ from those of a child for a kind parent, nor can I see in what respect the sense of fear for a cruel master differs from that of a child for a cruel parent. It is mere sentiment that ascribes to those of a child a higher source than the same passions in the dog--the dog could have loved or feared another master just as well; and filial love or fear would have reached out its tendrils just as far with all the ties of kindred blood removed. It has been said that one is able to assign a definite reason why, and that the other is a vague impulse; but I am too obtuse to understand how reason actuates to love, and instinct to a mere attachment. I cannot believe that in the essential and ultimate nature of these passions there can be shown any real difference. Whether it be reason or instinct in man, the affections of the lower animals are actuated by the same motives, governed by the same conditions, and guided by the same reasons as those of man. I shall not soon forget some of my monkey friends, and I am sure they will not forget me; for I see them sometimes after months of absence, and they usually recognise me at sight and show every sign of pleasure at my return.

CHAPTER X.

The Capuchin Vocabulary--What I have Found--What I Foresee in it.

Up to this time I have been able to determine with a fair degree of certainty nine words or sounds belonging to Capuchins, some of which sounds are so inflected as to have two or three different meanings, I think. The sound which I have translated food and found to have a much wider meaning, long perplexed me, because I found it used under so many conditions and had not been able to detect any difference of modulation. I find one form of this sound used for food in general, but when modulated in a certain way seems to specify the kind of food. I observed that this sound seemed to be a salutation or peacemaking term with them, which I attributed to the fact that food was the central thought of every monkey's life, and that consequently that word would naturally be the most important of his whole speech. During the past winter, I found that another modulation of this word expressed a wish to obtain a thing, and appeared to me to be almost equivalent to the verb "give," when used in the imperative mood, something like this, "Give me that." I have succeeded a great number of times, by the use of this word, in inducing McGinty to give me a part of his food, and on many occasions to hand me from his cage a ball, a club, or some such thing that I had given him to play with. Under suitable conditions, I could soon determine to what extent these inflections control their actions, but with the surroundings of a zoological garden the task is very difficult. However, I am quite satisfied that the sound which I have translated food is shaded by them into several kindred meanings.

The word "drink" appears to be more fixed, both in its form and meaning. I have not yet been able to detect any difference in the sound whether water, milk, or other liquids be desired; but this is quite natural, since they have but little variety in the things they drink.

[Sidenote: SOUNDS "WEATHER" AND "LOVE"]

The sound which I had thought meant "weather," or in some way alluded to the state of the weather, I am not sure how far that may be relied upon as a separate word. It was so closely connected to the speech of discontent or pain when made by little Dago, that I have not been able since to separate the sounds, and I finally abandoned it as a separate word; but reviewing my work, and recalling the peculiar conduct of this monkey and the conditions attending it, I believe it is safe to say that he had in mind the state of the weather.

The sound which I have translated "love" is only in the sense of firm and ardent friendship. The expressions of love between sexes I have not been able as yet to find with certainty. A few sounds, however, made under certain conditions, I have reason to believe bear upon this subject, but I am not yet ready to announce my opinions thereon.

The "alarm" sound, as I have translated it, has been described; but among the Capuchins I find three kindred words, quite unlike as mere sounds, but closely allied in meaning. The one just mentioned is used under the stress of great fear, or in case of assault. It is a shrill, piercing sound, very loud and very high in pitch. The second word, "e-c-g-k," used only to express apprehension, or as a warning of the approach of a thing they fear or do not like; and the last of these, which is a guttural whisper, is used merely to call attention to the approach of something which the monkey does not fear or dislike, which I have spelt "c-h-i."

I have referred elsewhere, without describing it, to the sound which Nellie used for calling, and which she employed when attempting to dissuade my wife from going out and leaving her alone. It is a peculiar sound, something like a whine, but very plaintive and suggestive. I cannot represent it in letters.

[Sidenote: THE CAPUCHIN TONGUE]

There are many sounds about which I am yet in doubt, and some shades of meaning are not clear, but these sounds described include the greater part of my knowledge of the Capuchin tongue, and I shall now proceed to the sounds of some of the other monkeys.

Standing on this frail bridge of speech, I see into that broad field of life and thought which lies beyond the confines of our care, and into which, through the gates that I have now unlocked, may soon be borne the sunshine of human intellect. What prophet now can foretell the relations which may yet obtain between the human race and those inferior forms which fill some place in the design, and execute some function in the economy of nature?

A knowledge of their language cannot injure man, and may conduce to the good of others, because it would lessen man's selfishness, widen his mercy, and restrain his cruelty. It would not place man more remote from his divinity, nor change the state of facts which now exist. Their speech is the only gateway to their minds, and through it we must pass if we would learn their secret thoughts and measure the distance from mind to mind.

CHAPTER XI.

The Word for Food in the Rhesus Dialect--The Rhesus Sound of Alarm--The Dialect of the White-face--Dolly Varden, "Uncle Remus," and others.

From a number of sounds uttered by the Rhesus monkeys, I finally selected the word which, for many reasons, I believed meant food, and was the equivalent in meaning to that word in the Capuchin tongue. The phonetic character of the words differs very widely. The sound uttered by the Rhesus, as nearly as I can represent it by letters, is "nqu-u-w." The "u" sound is about the same as in the Capuchin word, but on close examination with the phonograph it appears to be uttered in five syllables very slightly separated, while the ear only detects two.

One of the most unique of my experiments I made in Central Park, in the autumn of 1891. I secured a very fine phonograph record of the food sound of the Rhesus monkeys belonging to the Park. During the following night there arrived at the Park a shipment of Rhesus monkeys, just from their home in the east of Asia. There were seven of these new monkeys, three adult females and four babies, one of whom was left an orphan by the death of its mother in her passage across the ocean. At my request the superintendent had these monkeys stored in the vacant room in the upper story of the Old Armoury building. They had never seen the monkeys in Central Park, nor had they ever been brought near enough to the monkey-house for them to learn by any means that any other monkeys were about. About sunrise I repaired to this room, where I had my phonograph placed in order, and I enjoined those who were present, by special permission, not to do anything to attract the attention of the monkeys, nor under any condition to show them any food or anything to drink. Having arranged my phonograph, I delivered to them the sounds contained on my cylinder which I had recorded on the day preceding. Up to this time not a sound had been uttered by any inmate of the shipping cage. The instant my phonograph began to reproduce the record, the seven new monkeys began to answer vociferously. After having delivered this record to them, I gave them time to become quiet again. I showed them some carrots and apples, on seeing which they began to utter the same sounds which they had uttered before, and this time I secured a good record of their sounds to compare with the others.

[Sidenote: RHESUS MONKEYS]

The alarm-sound as given by the Rhesus is very energetic, but not so shrill nor sharp as that of the Capuchin, nor have I discovered more than one such sound. As they are not of a high order of intelligence, nor kindly disposed unless kept in fear, I have not given them a great amount of study, but their sounds come more closely to the range of the human voice than do the sounds of the Cebus, which I regard as the Caucasian of monkeys.

The Rhesus is not very intelligent, but when reared in captivity appears to be capable of some degree of domestication. The adult reared in a wild state shows many phases of vicious and uncongenial temper. When well cared for, they are rather hardy and undergo training quite well. They are not a handsome animal, being of a faded tan colour on the back, merged into a yellowish white on the less exposed parts. They have large cheek-pouches which, when not filled with food, allow the skin on the neck and jaws to hang in folds, which give them an appearance of extreme emaciation, and when full of food they are so distended as to present rather an unpleasant aspect.

The sounds which the Rhesus utters in anger are harsh and unmusical, while their sound for food is soft and sympathetic, and I have made a machine which imitates it quite well. The Rhesus belong to the genus Macacus, one of the oldest and largest of all Simian genera.

I have found the word in the dialect of the white-faced Cebus which corresponds in value to those sounds described in the dialects of the Capuchin and Rhesus monkeys meaning food, but I cannot give the faintest idea of the sound by any combination of letters, nor have I as yet devised any means by which I can imitate it. I recorded this sound on the phonograph more than a year ago, but only within the last few months have been able to tell its meaning.

[Sidenote: SOUND OF DANGER]

Another sound which is made by this species to express apprehension of remote danger, such as an approaching footstep or some unusual sound, I have also learned. It is very much the same phonetically as that sound which he utters in case of great and sudden alarm, but uttered with much less energy. It resembles slightly the alarm-sound of the Capuchin, but up to this time I have not been able to make a good record of it.

Another sound which is peculiar to this species I think is used as a kind of salutation or expression of friendship, which phonetically is quite unlike the corresponding sound in any other dialect that I have studied.

I must mention Dolly Varden, who belongs to this species, and with whom I was at one time on very warm terms of friendship. Dolly was very fond of me, and would laugh and play with me by the hour. Her laughter was very human-like, except that it was silent, and in all our play during the lapse of some weeks she never uttered a sound, not even so much as a growl, although I tried by every possible means to induce her to talk. It has occurred to me since that time that she may have been deaf and dumb, but I did not think of testing her on these points while I had an opportunity. It is not usual for monkeys to laugh in silence, although they frequently laugh aloud like human beings; but it is not a common thing for them to remain silent at all times and under all conditions. Dolly was good-natured, playful, and always showed every sign of pleasure at my visits.

[Sidenote: "UNCLE REMUS"]

In Central Park there is a monkey of this species which I call "Uncle Remus." He is quite fond of me, and, for my amusement, he always wants to whip a little baby monk in the same cage with him whenever I go to visit them. This species belongs to the same genus as the Capuchin, but they differ in mental calibre as widely as the Caucasian differs from the Negro; but in this case the colours are reversed. I have seen a few fairly intelligent white-faces and a great many very stupid Capuchins, but, to strike an average from a great number of each kind, they will be found very widely separated in brain power.

The white-faced Cebus always has a languid expression, and looks like some poor, decrepit old man, who has borne a great burden of care through a long life, and finds his toil and patience ill-requited and is now awaiting his last call. He always has a sad face, and looks as if his friends were false. His type of speech is very far inferior to that of the Capuchin, and I do not regard him as a good subject for my work.

[Sidenote: JIM AND THE MANGABY]

I have learned the food sound in the dialect of the sooty Mangaby, but I have not been able to record it sufficiently well to study; but it is one of the most peculiar sounds in the whole range of Simian speech. The phonetic elements are nearly like "wuh-uh-uh," but the manner in which it is delivered is very singular. It appears to be intermixed with a peculiar clucking sound, and each sound seems independent of the other, although so closely joined in their utterance as to sound almost like they were uttered simultaneously by separate means. It is a deep guttural, below the middle pitch of the human voice, while the clucking element appears much higher in pitch, and the whole sound is marked with a strong tremolo effect. The syllables are uttered in rapid succession, and this peculiar sound under different conditions is uttered in at least three different degrees of pitch about an octave apart, but the contour appears to me the same in each. This species talks but little, is very shy, makes few friends, and is afraid of the phonograph; hence I have never been able to make a good record of its voice. I was cultivating the friendship of Jim, who recently died in Central Park, and we were getting on the best of terms; but the little Mangaby that survives him is very shy and suspicious. Immediately after Jim's death, however, when I would visit the Garden, she would always jump on the perch and take the same position that Jim had occupied whenever I would feed him. During his lifetime, she always kept her distance and never would take anything out of my hand, because she was afraid of him; but as soon as he was out of the way she assumed his place, and would utter the same sound that he had uttered at my approach. She evidently was aware of the fact that Jim and I were friends, that I always gave him something good to eat at that particular place in the cage, and that he always sat in a certain position when I gave it to him. I do not regard this species as very intelligent, nor their language as being of a high type; but they have a very human-like face, almost without hair, and very large and expressive eyes. They abound in West Africa, and have been colonised with success in the island of Mauritius; they are not very common in captivity, but much more so than some other species of less interest.

CHAPTER XII.

Atelles or Spider Monkeys--The Common Macaque--Java Monkeys, and what they say--A Happy Family.

I have caught one sound from the spider monkey by which I have been able to attract the attention of others of the same species, but I am as yet uncertain about its meaning. I do not believe that it has any reference to food; but I think perhaps it is a term of friendship, or a sound of endearment. One reason for this belief is, that I have heard it used on several occasions when a monkey of this kind would see its image in a mirror. I have used the sound in

Washington, Philadelphia, and Atlanta, and induced the monkey addressed to respond to it and come to me. I almost concluded at one time that this species was nearly dumb, until I saw one enraged by a green monkey that occupied an adjoining cage. On this occasion she raised her voice to an extremely high pitch, and uttered a sound having great volume and significance. This she repeated several times, and it was the first time I had ever seen a spider monkey show any sign of resentment. On another occasion, where this same specimen saw a brilliant peacock near the window by her cage, the sounds which she made at that strange object were loud, clear, and varied.

I have read with surprise an account of a spider monkey which Dr. Gardner had with him in his travels through South America. He describes it as the most intelligent of all monkeys, but I cannot believe that his experience with monkeys was sufficient to rank him as an authority on that subject. I do not pretend, however, to know all that there is to be known concerning this species, but so far as my study of them goes they scarcely laugh, cry, or show any sign of emotion. They do not usually resent anything; thus they are harmless and timid. Their long, lean, half-clad limbs look like the ghost of poverty, and their slow, cautious movements like decrepitude begging alms. They would be objects of pity if they only had sense enough to know how Nature has slighted them.

[Sidenote: "JESS"]

I have recently received a letter from Mr. A. E. McCall, of Bath, New York, enclosing a photograph of a monkey of this kind, by the name of "Jess." The gentleman tells me that he has been giving some time to the study of the actions and language of this monkey, and assures me that it is very docile, and follows him like a dog, and kindly offers to make such experiments with it as I may suggest, by which to aid me in the pursuit of my own researches, and I shall take advantage of his kind offer.

I am aware that there are exceptions to all rules, and I am not disposed to deprive the spider monkey of the place he may deserve in the scale of Simian life by reason of his intellect or speech; but as this book is a record of what I know, and not what I have heard of, I shall for the present be compelled to place the spider monkey very far down in the scale of intellect and speech.

The common Macaque is a strong, well-built monkey, of a dark grey colour, with a short stubby tail. He has but few friends, and at times appears to regret having any at all. He is quite active, energetic, and aggressive. He endures captivity well, but as a rule never becomes quite tame or trustworthy. His speech is of a low type, but he has a very singular expression of the mouth, which seems to indicate friendship. In fact, there are several different species of the genus Macacus that use this peculiar movement of the lips. They thrust the head forward and lower it slightly, and in this position work their lips as if talking with the greatest possible energy, but without uttering a sound. They do not do this for food, but I have seen them do it to their image in the glass, and have had them do so with me a great number of times. I have been told by some that this is meant as a sign of anger or assault, but my own observations tend to attribute to it exactly the reverse of this meaning. Occasionally, when I have offered them food, I have observed them do this; but I do not think it referred to the food, unless it was intended as a vote of thanks. The first monkey whose voice I ever captured on the phonograph belonged to this tribe; he is still in the Washington collection, and bears the name of "Prince," under which name he may go down to history as the first monkey whose speech was ever recorded. But whatever his fame may become on that account, I do not think he will ever justly obtain the reputation of being an amiable monkey.

[Sidenote: JAVA MONKEYS]

Among the Java monkeys are several varieties which make very good pets. They show a fair degree of intelligence and docility, and are not generally very vicious. I have not succeeded in making any very good records of these monkeys, although I have observed, without the aid of the phonograph, that they have one or two very distinct and well-marked sounds. I have not up to this time attempted to differentiate their sounds, but in a general way have interpreted the meanings of one or two groups of them, especially those of a friendly character. I may with propriety remark here, that in all the different tongues of monkeys there appear to be certain words which are much more significant, of a much better phonetic type, than the others, and occur much more frequently among their sounds. This appears to be true of the speech or sounds of all the lower animals.

In a former chapter I have described the happy little family in Central Park, which consisted of the five little brown cousins, only a few months ago; but death has reduced their number to two. In this connection I shall mention a very important fact concerning the use of the natural senses of these animals. I have several times been assured that monkeys depended more upon their sense of smell than upon that of sight as a means of recognition, and that in this respect they were very much like the canines. I have made frequent tests of the power of their senses, and am prepared to say with certainty that such is not the case. When I visit the Park, I frequently enter at Sixty-fourth Street and Fifth Avenue, at which place there is a flight of stairs leading from the street down to a large plazza in front of the Old Armoury; and something more than a hundred feet from the foot of the stairway, and nearly at right angles to it, is a window opening into the monkey-house by the cage occupied by these particular monkeys. When I descend the stairway and come within view of this window, they frequently see me as I reach the plazza, and the keeper always knows of my approach by the conduct of the monkeys, who recognise me the instant I come in sight at that distance. At other times I have approached the house from another direction, and come within a few feet of their cage, where I have stood for some time, in order to ascertain whether they were aware of my presence; and on a few occasions have slipped into the house with the crowd, and they did not detect my presence except by sight. It is evident, if they depended upon the sense of smell, that they would have discovered my presence when so near them, although they could not see me. But no matter what the condition of the weather, or how many people are present, the instant one of them sees me he spreads the news, and every inmate of the cage rushes to the window and begins to scream at the top of his voice. If their sense of smell was such as to enable them to detect my presence as a dog would, it is reasonable also that the monkey which possessed the most sensitive organs would have been the first to detect it in each case; whereas, sometimes one monkey, and sometimes another, made the discovery. It is my belief, however, that their sense of smell is much more acute than that of man, but far less so than that of most other animals, especially the dog. [Sidenote: HEARING VERY DELICATE] The sense of hearing in these animals is very delicate, as may be seen from the account of Nellie discovering my footsteps on the lower stairway, and as I have witnessed in scores of other cases. The same is true also of their sight;

their eyes are like a photo-camera, nothing ever escapes them. I think their organs of taste are also quite sensitive, as I have made some tests from time to time, and find them very hard to deceive. The sense of touch, which is rather obtuse in most animals, is much more acute in these. I have frequently interlaced my fingers with those of some person whom they dislike, and extending the hand towards them, they rarely make a mistake by getting hold of the wrong finger, and yet it has frequently occurred that they could not see the hands at all, and had to depend alone upon the sense of touch. In cases where the hands were very nearly the same size they were not able to select the fingers so readily, but where a lady's hand was used, or that of a boy, the selection was made without hesitancy and without error. I have tried this experiment a great many times with a view to ascertaining to some extent the delicacy of their sense of touch. Another fact that I may mention is, that they do not habitually smell articles of food or other things given to them; but they depend chiefly upon their sight for finding and their taste for choosing their food. My opinion is, that the sense of smell does not play an important part in these affairs. I may add, too, that, in the Cebus, his tail is perhaps the most sensitive organ of touch, although it is not used in this capacity to any great extent. He is generally very watchful over this useful member, because it serves him in so many ways, and I think perhaps it is safe to say that the tail is the last part of the monkey that ever becomes tame.

CHAPTER XIII.

The Extent of my Experiments--Apes and Baboons--Miscellaneous Records of Sound--The Vocal Index.

In quest of the great secret of speech, I have pursued my investigations chiefly in the direction of learning one tongue, but incidentally I have made many detours, and I have recorded the sounds of many other forms of the animal kingdom, besides primates. I have examined the phonation of lions, tigers, leopards, cats, dogs, birds of many kinds, and the human voice in speech, music, and laughter. Besides these, I have examined various musical sounds, especially of the pipe and whistle kinds.

More than a year ago I made some splendid records of the sounds of the two chimpanzees in the Cincinnati collection. I have not had the opportunity to study these apes themselves, as I desired to do, since they are kept so closely confined in a glass house, and for ever under the eye of their keeper, which conditions are not favourable to the best results. I am not prepared therefore to give much detail concerning their speech; but from a careful study of one cylinder containing a record of their sounds, I was able to discern as many as seven different phones, all of which come within the scope of the human vocal organs. I learned one of these sounds, and on a subsequent visit to Cincinnati I succeeded in attracting the attention of the female, and eliciting from her a response. She would come to the lattice door of the inner cage by which I was standing, and when I would utter the sound she would press her face against the door of the cage and answer it with a like sound. The male, however, did not appear to notice it with any degree of concern. I have no idea what the sound meant, and my opportunities have not been such that I could translate it with the remotest degree of certainty. [Sidenote: STUDIES IN TROPICAL AFRICA] These apes will be one of the chief objects of my studies in tropical Africa, as I believe them to possess a higher type of speech even than the gorilla. In this opinion, which I reached from the study of other sounds and the types of skull to which they belonged, I am not alone: Mr. Paul Du Chaillu, Mr. E. J. Glave, and others who have seen both of these apes in their native habitat, agree with me on this point. I am aware that this view is not in strict accord with that of Professor Huxley, who assigns the gorilla the highest place next to man in the order of Nature, and the chimpanzee next below him. I shall not here attempt to discuss the question with so high an authority, and I must confess that the vocal index is not yet so well defined that it may be relied upon in classifying apes. One aim I have in view is to study the gorilla and chimpanzee side by side in their native wilds, and to record, if possible, the sounds of their voices in a wild state. From the study of the sounds I have made, I feel confident that all the vocal sounds made by these apes may be uttered by the human vocal organs.

Some months ago I made a record of the voice of the great Anubis baboon, in Philadelphia. I did not expect to find in him an elevated type of speech; but my purpose was to compare it with other Simian sounds, to see if I could not establish a series of steps in the quality of vocal sounds which would coincide with certain other characters. I had found by the study of certain cranial forms that certain vocal types conformed to certain skulls, and were as much a conformation thereof as are the cerebral hemispheres. I then believed, and have had no cause since to recede from it, that the vocal powers were correctly measured by the gnathic index; that the mind and voice were commensurate; and that as the cranio-facial angle widens the voice degrades in quality and scope. In man, I find the highest vocal type, and just as we descend in the cranial scale, the vocal type descends into sounds less flexible, less capable, and less musical. These deductions apply only to mammals; among birds, insects, &c., a different order may prevail.

[Sidenote: RECORDS OF LIONS]

The records of the lions show some strange features in the construction of sound; and when analysed on the phonograph present some novel effects. The sound as a whole appears to be broken into broad waves or pulsations; but on analysing it the fundamental tones somewhat resemble the sounds produced by drawing a mallet rapidly across the keyboard of a xylophone, and are characterised by a peculiar resonance something like the tremulous vibrations of a thin glass containing a small quantity of water. Each of these separate fundamental sounds, or sound units as they appear to be, can be further reduced to still smaller vibrations; and the result suggests that the fundamental sounds themselves are an aggregation of smaller vibrations. I have not as yet been able to compare the notes one by one with the scale of the xylophone in order to ascertain whether or not they obey the laws of sound upon which is founded the chromatic scale of music. The lion makes only a small number of different sounds, nearly of the same pitch. I have not analysed the vocal sounds of the other felines to ascertain to what extent they coincide with those of the lion; but his appear to be somewhat unlike any other sounds which I have examined.

Among the few sounds of birds which I have analysed, I may mention the Trumpeter Crane. I have made one record of this bird which was sufficiently loud to enable me to obtain some idea of the character of the sound. I am in doubt as to what the real mode of producing this sound is. The volume of sound evidently comes from the mouth of the bird; but while in the act of making it, he appears to bring the whole body into use, even the feathers appear to take some part in its production, and the whole frame of the bird vibrates in the act. The record which I have shows some resemblance, on analysis, to the sound made by the lion; but it is not sufficiently strong to admit of analysing the sound units or fundamental sounds.

[Sidenote: DIFFERENCE IN PHONES OF GENERA]

From the many sounds that I have analysed, it appears to me that there is a difference in the phones of all different genera, and that the phonetic basis of human speech more closely resembles that of the Simian than any other sounds; but I wish to be understood distinctly not to offer this in evidence to establish any physical, mental, or phonetic affinity between mankind and Simians. I merely state the facts from which all theorists may deduce their own conclusions.

CHAPTER XIV.

Monkeys and the Mirror--Some of their Antics--Baby Macaque and her Papa--Some other Monkeys.

I have incidentally mentioned elsewhere the use of the mirror in some of my experiments, but I have not described in detail how it affected various monkeys. Of course, it does not always affect the same monkey in the same way at different times, nor does it affect all monkeys of the same species in exactly the same way, and therefore I cannot deduce a rule from my experiments by which the species can be determined by its conduct before the glass.

[Sidenote: PUCK AND NELLIE WITH MIRROR]

When Puck saw himself in the mirror he undoubtedly mistook the image for another monkey, to which he would talk more freely than he would to the sounds made by the phonograph. He would frequently caress the image, and show signs of friendship; at the same time he was very timid and retiring.

Nellie would chatter to herself in the mirror, and seemed never to tire of looking at that beautiful monkey she saw there, and I do not think the propensity could be accounted for merely by her sex. I do not think she ever quite understood where that monkey was concealed, and the scores of times in a day that she would turn the glass around was evidence that she never fully despaired of finding it.

I accidentally dropped a small mirror one day by the cage in which there was a green monkey. The glass was broken into many small pieces. Quick as thought, the green monkey thrust her arm through the bars, grabbed the largest piece, and got it into her cage before I was fully aware of what she was trying to do. The fragment was about an inch wide by an inch and a half long. She caught a glimpse of herself in the glass, and her conduct was more like that of a crazy monkey than anything I can compare it to. She peeped into the fragment of the mirror, which she seemed to regard as a hole in something which separated her from another monkey. She held it up over her head at arm's-length, laid it down on the floor, held it against the wall, and twisted herself into every pose to get a better peep at that mysterious monkey on the other side of something, she could not tell what. When the glass was reversed, she seemed much perplexed, and would sometimes jump high off the floor, and turn herself entirely around, as if to untangle the mystery. Then again she would discover the right side of the glass, and would go through these antics again. Several times while holding it against the wall she would put her eyes close up against the glass, just as she would to a knot-hole in the wall. I tried in vain for some time to get the glass away from her lest she might injure herself with it, but only succeeded after considerable labour and through the help of her keeper.

[Sidenote: McGINTY'S DELIGHT WITH MIRROR]

McGinty always tries to find the image behind the glass. He reaches his little black hand as far as he can around behind it, peeps over and under it, pecks on the glass with his fingers, kisses and caresses it, and grins at it with infinite delight. He often tries to turn the glass around to look on the back of it, and when he finds no monkey there he works his eyebrows as if perplexed, and utters a sound which reminds me of a child under similar circumstances saying "gone" when in play something is concealed from it to make the child believe it is lost. Then he will suddenly turn the glass around again, as if the thought had just occurred to him, and when he again discovers the image, he will laugh, chatter, peep and peck at the glass, as if to say "There it is, there it is!" But, like all other monkeys, he does not quite understand where that monkey conceals itself when he peeps over the glass.

Mickie does not appear to enjoy the sight of himself in the glass. He always looks at it earnestly but doubtfully, and utters a low sound in a kind of undertone, frowns and scowls as though he regarded the new monkey as an intruder. He rarely talks to the image only with this low, muttering sound, and never tries to find it by reaching his hand behind the glass or making any other investigation. Mickie, however, has been very much petted, in consequence of which he is very selfish, just as children become under like treatment.

Little Nemo always looked at himself in the glass in the most inquisitive and respectful manner, without ever winking an eye or betraying any sign of emotion, except that he would caress the image in the glass over and over again by pressing his lips to it in perfect silence. Indeed, his conduct would suggest to you that he regarded the image as a portrait of some dear departed one, which awoke the tender memories of the past and filled the heart too full for utterance. His sedate manners were very becoming.

Dodo always appeared to be afraid of the image. She would merely take a peep and turn away. She would sometimes utter a single sound, but rarely touched her mouth to the glass, and never felt behind it for the other monkey. This, perhaps, was due to the fact that she was afraid of some of the other inmates of the cage, and I do not think that she desired the colony increased.

Nigger always showed great interest in the mirror when left alone, but when the other monkeys would crowd around to peep into the glass he would always leave to avoid trouble with them.

[Sidenote: "UNCLE REMUS," THE WHITE-FACE]

"Uncle Remus," the white-face, always goes through a series of facial contortions with the gravity of a rural judge. He will look into the glass, and then at me, as if to say "Where did you get that monkey?"

The little baby Macaque, who was born in Central Park, tries to engage the image in a romp, reaches for it in the glass, clucks, jumps playfully to her perch, and looks back to see if the image follows; then she will return to the glass, and try again to induce the little ghost to join her in her play. Again, she will spring to her perch, looking back, but does not understand why it will not join her. During all this, the baby's father, a sedate old Macaque, looks on with suspicion and a scowl, and on a few occasions has pulled the baby away from the glass, as if he knew that there was something wrong, and expressed his opinion in a low, ominous growl. He reminds me at times of some people whom I have seen that look very wise, and intimate by their conduct that they know something.

Another little Macaque makes the most indescribable faces, and works her lips in that peculiar fashion which I have elsewhere described, but she does not utter one sound. She merely looks in silence, and never tries to find the monkey concealed behind the glass.

[Sidenote: THE SPIDER MONKEY]

The spider monkey is a study worthy of great minds. When shown her image in the glass, she takes her seat on the floor, crosses her legs, and fixes herself as if she expected to spend the day there. She will then look into the glass and utter a low sound, and begin to reach out her long arms in search of the other monkey. It is surprising to see how she will adjust her reach as you change positions with the glass. Of course, as you remove the mirror from her the image is removed accordingly, and she extends or contracts her reach to suit that distance. This is not, however, an evidence of her mathematical skill, since to her mind the image is doubtless a real thing, and she is governed by the same instinct or judgment in reaching for it as she would be if it were real. More than any other, the spider monkey seems to admire herself in the glass; notwithstanding she is about the homeliest of all the Simian tribes, yet she will sit for hours in almost perfect silence, and gaze upon her image.

CHAPTER XV.

Man and Ape--Their Physical Relations--Their Mental Relations--Evolution was the Means--Who was the Progenitor of the Ape?--The Scale of Life.

If we could free our hands from the manacles of tradition and stand aloof from our prejudices, and look the stern facts in the face, we should be compelled to admit that between man and ape there is such a unity of design, structure and function, that we dare not in the light of reason deny to the ape that rank in Nature to which he is assigned by virtue of these facts. Physiologically, there is no hiatus between man and ape which may not be spanned by such evidence as would be admitted under the strictest rules of interpretation. We may briefly compare these two creatures in a broad and general way, so that the unscientific and casual reader may comprehend.

The skeleton of man is only the polished structure of which that of the ape is the rough model. The identity of the two, part by part, is as much the same as the light sulky is the outgrowth of the massive framework of the old-time cart. Whether man and ape are related by any ties of blood or not, it is evident that they were modelled on the same plan, provided with the same means, and designed for like purposes, whatever they may be. The organs of sensation and the functions which they discharge are the same in both, and the same external forces addressing themselves thereto produce the same results. I do not mean to say that the same organ in each is developed in the same degree as that in the other, for this is not the case even in different individuals of the same kind. In the muscular system of the one is found an exact duplicate of the other, except in such slight changes of model as will better adapt the parts to those conditions of life under which the animal having them may be placed, and through the whole physical structure of both we find that unity of part and purpose in structure and function, in bone, muscle, nerve, and brain. It has been shown beyond a reasonable doubt that the brain in the higher races of mankind has reached its present form through a series of changes which are constant and definite; and this organ in the lower types of man resembles more that of the ape than does the same organ in the higher types of man; and by a method of deduction, such as we use to determine the height of a tree or the width of a stream by the length of a shadow, we find that the fiducial lines which bound the planes in the perspective of man's cerebral growth, likewise embrace those of the ape. While it is a fact that the mind of man so far transcends that of the ape, it is also a fact that in reaching this condition it has passed through such planes as those now occupied by the ape. The physical changes of man's brain do not appear to keep pace with the growth of his mind. This may be a paradox, but the evidence upon which it rests is ample to sustain it.

I do not pretend to know whether man was evolved from ape, or ape from man; whether they are congenetic products of a common authorship, or the masterpieces of two rival authors; but I cannot see in what respect man's identity would be affected, whatever may be the case. If it be shown that man descended from the ape, it does not change the facts which have existed from the beginning, nor does it change the destiny to which he is assigned. If it can be shown that apes descended from man, it does not leave upon man the censure for this degeneracy. If man has risen from the low plane of brutehood which the ape now occupies, has scaled the barriers which now separate him from apes, and has climbed to the divine heights of mental and moral manhood, the ape deserves no praise for this. On the other hand, if apes have fallen from the state of man, have wandered so far from the gates of light, and are now wandering in the twilight of intellect and degradation, it is no reproach to man; and while I shall not sit in judgment in the cause, nor testify on either side, I am willing to accept whatever verdict may be founded on the real facts, and I shall not appeal therefrom. But I shall not allow my prejudice to conceal the truth, whenever it is shown to me. It is always acceptable to my mind, and, stripped of all sophistry and oblique conditions, it would appear the same to every mind.

That evolution is the mode by which the world was peopled, there is little doubt, but there are many details yet unsettled as to the manner in which this was effected. I cannot regard the matter as proven beyond appeal that man has come from any antecedent type that was not man, nor yet do I deny that such may be the case; but I do deny that the broad chasm which separates man from other primates cannot be crossed on the bridge of speech; and while this does not prove their identity or common origin, it does show that Nature did not intend that either one should monopolise any gift which she had to bestow. It is as reasonable to believe that man has always occupied a sphere of life apart from that of apes, as to believe that apes have occupied a sphere of life apart from birds, except that the distance from centre to centre is greater between birds and apes than that distance between apes and man. So far as any fossil proofs contribute to our knowledge, we find no point at which the line is crossed in either case; and the earliest traces of man's physiological history find him distinctly man, and this history reaches back on meagre evidence many, many centuries before historic time. Among these earlier remains of man, we find no fossils of the Simian type to show that he existed at that time; but at a somewhat later period we find some remnants of the Simian type in deposits of Southern Europe; but they are of the smaller tribes, and have been assigned to the Macacus. We cannot trace the history of this genus from that to the present time to ascertain whether they were the progenitors of apes or not; but between this type and that of apes the hiatus is as broad as that which intervenes between the ape and man.

That somewhere in the lapse of time all genera began, admits of no debate; and by inversion it is plain that all generic outlines must focus at the point from which they first diverged, and such an operation does not indicate that man and Simian have ever been more closely allied than they are at the present time; but the evidence is clear that man has been evolved from a lower plane than he now occupies. The inference may be safely applied to apes, as progress is the universal law of life.

The question has been asked, "Who is the progenitor of man?" The solution of this problem has engaged the most profound minds of modern time. If it be said in reply that apes were the progenitors of man, the question then arises, "Who was the progenitor of the ape?" If it be said that man and ape had a common progenitor, a like question arises, and it becomes necessary to connect all types allied to each other as these two types are physically allied. If man is the climax of a great scheme in Nature by which one type is gradually transformed into another, we must descend the scale of life by crossing the chasm which lies between mankind and apes, another lying between the apes and monkeys, another between the monkeys and baboons, another between the baboons and lemurs, and yet another between the lemurs and the lemuroids, and thus from form to form like islands in the great sea of life. From man to infinity the question constantly recurs, and over each hiatus must be built a separate bridge.

[Sidenote: DARWIN'S PROFOUND WORK]

Darwin has given to the world the most profound and conscientious work, and from the chaos and confusion of human ignorance and bigotry has erected the most sublime monuments of thought and truth. It does not detract from his character and honesty, nor lessen the value of his labours, to admit that he may have been mistaken in some conclusions which he deduced from the great store of facts at his command.

It is not the purpose of this work, however, to enter into a discussion of any theory aside from speech and its possible origin and growth, but all subjects pertaining to life, thought, and the modes of living and thinking, must contribute in some degree to a clear understanding of the subject in hand.

It has been a matter of surprise to me that so careful and observant a man as Mr. Darwin should have so nearly omitted the question of speech from a work of such ample scope, such minute detail, and such infinite care as characterises the "Descent of Man," and such like works. But science will cheerfully forgive an error, and pardon the sin of omission in one who has given to the world so much good.

CHAPTER XVI.

The Faculty of Thought--Emotion and Thought--Instinct and Reason--Monkeys Reason--Some Examples.

The study of biology has revealed many facts which conspire to show that the incipient forms of animal and vegetable life are the same in those two great kingdoms; and parallel with this fact, I think it can be shown that the faculty of expression goes hand in hand with life. And why should not this be the case? From the standpoint of religion, I cannot see why the bounty of God should not be equal to such a gift, nor can I conceive of a more sublime act of universal justice than that all things endowed with thought, however feeble, should be endowed with the power of expressing it. From the standpoint of evolution, I cannot understand by what rule Nature would have worked to develop the emotions, sensations, and faculties alike in all these various forms, and make this one exception in the case of speech. It does not seem in keeping with her laws. From the standpoint of chance, I cannot see why such an accident might not have occurred at some other point in the scale of life, or why such anomalies are not more frequent. Man appears to be the only one. From any point of view we take, it does not seem consistent with other facts. All other primates think and feel, and live and die under like conditions and on like terms with man; then why should he alone possess the gift of speech?

I confess that such an inference is not evidence, however logical; but I have many facts to offer in proof that speech is not possessed by man alone. It is quite difficult to draw the line at any given point between the process of thought and those phenomena we call emotions. They merge into and blend with each other like the colours in light, and in like manner the faculty of speech, receding through the various modes of expression, is for ever lost in the haze and distance of desire. The faculty of reason blends into thought like the water of a bay blends into the open sea; there is nowhere a positive line dividing them. When we are in the midst of one we point to the other, and say, "There it is;" but we cannot say at what exact point we pass out of one into the other.

[Sidenote: THE POWER OF REASONING]

To reason is to think methodically and to judge from attending facts. When a monkey examines the situation and acts in accordance with the facts, doing a certain thing with the evident purpose of accomplishing a certain end, in what respect is this not reason? When a monkey remembers a thing which has passed and anticipates a thing which is to come; when he has learned a thing by experience which he avoids through memory and the apprehension of its recurrence, is it instinct that guides his conduct? When a monkey shows clearly by his actions that he is aware of the relation between cause and effect, and acts in accordance therewith, is it instinct or reason that guides him? If there be a point in the order of Nature where reason became an acquired faculty, it is somewhere far below the plane occupied by monkeys. Their power of reasoning is far inferior to that of man, but not more so than their power of thinking and expression; but a faculty does not lose its identity by reason of its feebleness. When the same causes under the same conditions prompt man and ape alike to do the same act in the same way, looking forward to the same results, I cannot understand why the motive of the one should be called reason, and that of the other called instinct. Scholars have tried so hard to keep the peace between theology and themselves, that they have explained things in accordance with accepted belief in order that they might not incur the charge of heresy. To this end they have reconciled the two extremes by ignoring the means, and making a distinction without a difference on which to found it.

Whatever may be the intrinsic difference between reason and instinct, it is evident to my mind that the same motives actuate both man and ape in the same way, but not to the same extent. I am aware that many acts performed by Simians are meaningless to them and done without a well-defined motive. The strong physical resemblance between man and ape often causes one to attach more importance to the act than it really justifies. In many cases the same act performed by some other animal less like man would scarcely be noticed. To teach an ape or monkey to eat with knife, fork, cup and spoon, to use a napkin and chair, or such like feats, does not indicate to my mind a high order of reason; nor it is safe to judge the mental status of these creatures from such data. When he is placed under new conditions and committed to his own resources, we are then better able to judge by his conduct whether he is actuated by reason or not.

[Sidenote: CONNECTING CAUSES AND EFFECTS]

In any simple act where a monkey can see the cause connected with, and closely followed by, the effect, he is actuated by reason, and while he may not be able to explain to his own mind a remote or complex cause but simply accepts the fact, it does not make the act any less rational in a monkey than the same act would be in man where he fails to grasp the ultimate cause. The difference is that man is able to trace the connecting causes and effects through a longer series than a monkey can. Man assigns a more definite reason for his acts than a monkey can; but it is also true that one man may assign a more definite reason for his acts than another man can for his when prompted by the same motives to the same act.

The processes, motives, acts and results are the same with man and ape; the degree to which they reason differs, but the kind of reason in both cases is the same.

I shall here relate some instances in my experience and leave the reader to judge whether reason or instinct guided the acts of the monkeys as I shall detail them in the next few paragraphs. It will be remembered that these were new conditions under which the monkeys acted.

I taught Nellie to drink milk from a bottle with a rubber nipple. While I would hold the bottle, it was easy for her to secure the milk; but when she undertook it alone, she utterly failed. The thing which puzzled her was how to get the milk to come up to her end of the bottle. She turned it in every way, and held it in every position that she could think of, but the milk always kept at the other end of the bottle. She would throw the bottle down in despair, and when she saw the milk flow to the end having the nipple, she would go back and pick it up, and try it again. Poor Nellie worried her little head over this, and again abandoned it in despair. While trying to solve the mystery, she discovered a new trick. While the bottle was partly inverted she caught hold of the nipple, and squeezed it. By this means she accidentally spurted the milk into the faces of some ladies who were watching her. This afforded her so much fun that she could scarcely be restrained, and while she remained with me she remembered this funny trick, and never failed to perform it when she was allowed to do so. It was no trouble for her to connect the immediate effect to the immediate cause. But she could not for a long time understand that the position of the bottle or the location of the milk in it had anything to do with the trick. In the course of time, however, she learned to hold the bottle so that she could drink the milk, and she also discovered that it had to be held in a certain position in order to play her amusing trick.

Another instance was in the case of a little monkey, heretofore described by the name of Jennie. When you would throw a nut, just out of her reach, she would take a stick which had a nail in the end, and rake the nut to her. She never took the wrong end of the stick, and never placed the nail on the wrong side of the nut. Her master assured me that she had not been taught this, but had found the stick and applied it to this use. When she did not want any one to play with her or handle her, she would coil her chain up and sit down on it to keep any one from taking hold of it.

It is not an uncommon thing for monkeys to discover the means by which their cage is kept fastened, and they have frequently been known to untie a knot in a rope or chain, and thus release themselves. I have known a monkey that learned to reach its hand through the meshes of the cage, and withdraw the pin which fastened the hasp and thus open the door and get out. The keeper substituted a small wire, which he twisted three or four times in order that it could not be released. The monkey realised that the wire performed the duties of the pin and prevented the door from opening. He also knew that the wire was twisted and that this was the reason he could not remove it. I have seen him put his hand through the meshes of the cage, catch the loose end of the wire and turn it as though he was turning a crank. He evidently knew that the twist in the wire was made by such a motion and his purpose was to untwist it, but so far as I know he never succeeded in doing so. I have frequently seen a monkey gather up his chain and measure his distance from where he stood to the point at which he expected to alight, with the skill and accuracy of an engineer.

A gentleman of my acquaintance assured me recently that during his sojourn of two years in the Island of Sumatra, he had in his service a large orang. This ape did many chores about the place, and performed many simple duties as well as the other domestics did.

On one occasion, this ape was induced to go aboard a steamer which lay in the harbour. The purpose was to kidnap him and carry him to Europe. Either through fear, instinct, reason, or some other cause, this ape jumped overboard and swam ashore, although he was naturally afraid of water. From that time on to the end of the gentleman's residence there, he assures me that whenever a steamer made its appearance in the harbour, the ape would take flight to the forest, where he would stay as long as the vessel remained in sight. He was seen from time to time, but could not be induced to return to the house until the vessel had departed.

A few years ago, I saw on board the United States receiving ship Franklin, a bright little monkey which was kept chained in a temporary workshop built on the gun-deck. Her chain was just long enough to allow her to reach the stove. The day was pleasant outside, but in the shade a trifle chilly. The little monk descended from the sill on which she usually sat and carefully felt the top of the stove with her hands. Finding it slightly warm, although the fire had died out, she mounted the stove and laid the side of her head on the warm surface. She would turn first one cheek and then the other, and continued rubbing the stove with her hands. Not finding it warm enough, she jumped down on the floor, opened the stove door with her hand, and slammed it two or three times. She then picked up a stick of wood lying within reach, and tried to lift it to the stove. The stick was too heavy for her to handle, so she would lift up one end of it and drop it heavily on the floor with the evident purpose of attracting the attention of her master. Again she would open and slam the door, lift up the end of the stick and drop it, and utter a peculiar sound, showing in every possible way that she wanted a fire. She finally picked up a small stick and stuck the end of it into the ashes in the front of the stove. She knew that it was necessary to put the wood into the stove; she knew where to put it in, and, while she could not do it herself, she knew who could put it in. Her master told me that she would gather up the shavings from the floor when they came within her reach and pile them up by the stove. He also told me that he frequently gave her a lighted match when he had prepared the fuel for building a fire, and that she would touch the match to the shavings and start the fire. She never ventured to get on the stove without first examining it to ascertain how hot it was.

Another feat which she performed was to try to remove some tar from the cup in which he gave her water and milk. The cup had been lined with tar as a sanitary measure to prevent consumption, and she was aware that the tar imparted an unpleasant taste and odour, hence she tried very hard to remove it from the cup. Was this instinct?

CHAPTER XVII.

Speech Defined--The True Nature of Speech--The Use of Speech--The Limitations of Speech.

[Sidenote: SPEECH DEFINED]

What is speech? I shall endeavour to define it in such terms as will relieve it of ambiguity, and deal with it as a known quantity in the problems of mental commerce. Speech is that form of materialised thought which is confined to oral sounds, when they are designed to convey a definite idea from mind to mind. It is, therefore, only one mode of expressing thought, and to come within the limits of speech, the sounds must be voluntary, have fixed values, and be intended to suggest to another mind a certain idea, or group of ideas, more or less complex. The idea is one factor, and sound the other, and the two conjointly constitute speech. The empty sounds alone, however modulated, having no integral value, cannot be speech, nor can the concept unexpressed be speech. Separately, the one would be noise, and the other would be thought; and they only become speech when the thought is expressed in oral sounds. Sounds which only express emotion are not speech, as emotion is not thought, although it is frequently attended by thought, and is a cause of which thought is the effect. Music expresses emotion by means of sounds, but they are not speech; and even though the sounds which express them may impart a like emotion to the hearer, they are not speech.

The sounds which express crying, sighing, or laughter, may indeed be a faint suggestion of speech, since we infer from them the state of the mind attending the emotions which produce them, yet they are not truly speech. To be regarded as speech, the expression must be preceded by consciousness, and the desire to make known to another the sensation by which the expression is actuated. As the impulse can only come from within, it appears that emotion is one source from which thought is evolved, and speech is the natural issue of thought. Desire gives rise to a class of thoughts having reference to the sensations which produce them, and such thoughts find expression in such sounds as may suggest supplying the want. As the wants of man have increased with his changing modes of life and thought, his speech has drawn upon the resources of sound to meet those increased demands for expression. It appears only reasonable to me that thought must precede in point of time and order any expression of thought, for thought is the motive of expression, and the expression of thought in oral sounds is speech. [Sidenote: NATURE OF SPEECH] Speech is not an invention, and therefore is not symbolic in its radical nature. True, that much that is symbolic has been added to it, and its bounds have been widened as men have risen in the scale of civil life, until our higher types of modern speech have departed so far from the natural modes of speech and first forms of expression, that we can rarely trace a single word to its ultimate source. And viewing it as we do from our present standpoint, it appears to be purely symbolic; but if that be so, then we must deny the first law of progress, and assign the origin of this faculty to that class of phenomena known as miracles, which once explained by increasing the mystery what we could not understand, and served at the same time to conceal the exact magnitude of our ignorance; but as we added little by little to our stock of knowledge, such phenomena were brought within the realm of our understanding, and to-day our children are familiar with the causes of many simple effects which our forefathers dared not attempt to solve, but reverently ascribed to the immediate influence of Divinity. If speech in its ultimate nature is symbolic, what must have been the condition of man before its invention, and how did he arrive at the first term or sound of speech? He did not invent sound nor the means of making it. He did not invent thought, the thing which speech expresses, and it is no more reasonable to believe that he invented speech than to believe that he invented the faculties of sight and hearing, which are certainly the natural products of his organic nature and environments. So far as I can find through the whole range of animal life, all forms of land mammals possess vocal organs which are developed in a degree corresponding to the condition of the brain, and seem to be in every instance as capable of producing and controlling sounds as the brain is of thinking: in other words, the power of expression is in perfect keeping with the power of thinking. From my acquaintance with the animal kingdom, it is my firm belief that all mammals possess the faculty of speech in a degree commensurate with their experience and needs, and that domestic animals have a somewhat higher type of speech than their wild progenitors. Why are all forms of mammals endowed with vocal organs? Why should Nature bestow on them these organs if not designed for use? One or the other of two conclusions seems inevitable. As a law of evolution and progress, all organs are imparted to animals for use and not for ornament. It seems consistent with what we know of Nature, to suppose that the vocal organs of these lower forms are being developed to meet a new requirement in the animal economy, or having once discharged some function necessary to the being and comfort of the animal, they are now lapsing into disuse and becoming atrophied. If they are in the course of development, it argues that the creature which possesses them must possess a rudimentary speech which is developing at a like rate into a higher type of speech. If they are in a state of decay or atrophy, it argues that the animal must have been able to speak at some former period, and that now, in losing the power of speech it is gradually losing the organ. In either case, the organs themselves would be in a state of development in harmony with the condition of the speech of the animal. [Sidenote: LIMITATIONS OF SPEECH] The function which speech discharges is the communication of ideas, and its growth must depend upon the extent of those ideas; and in all conditions of life, and in all forms of the animal kingdom, the uses of speech are confined to, and limited by the desires, thoughts, and concepts of those using it. Its extent is commensurate with requirement. To believe that there was a time in the history of the human race when man could not speak, is to destroy his identity as man, and the romance of the alalus could be justified from a scientific standpoint only as a compromise between the giants of science and superstition. Among the tribes of men whose modes of life are simple, whose wants are few, and whose knowledge is confined to their primitive condition, the number of words necessary to convey their thoughts is very limited. Among some savage races there are languages consisting of only a few hundred words at most, while as we rise in the scale of civil and domestic culture, languages become more copious and expressive as the wants become more numerous and the conditions of life more complex. As we descend from man to the lower animals, we find the types of speech degenerate just in proportion as we descend in the mental and moral plane, but it does not lose its identity as speech. Through the whole animal kingdom from man to protozoa, types of speech differ as do the physical types to which they belong. But as the same vital processes are found throughout the whole circle of life, so the same phonetic basis is found through the whole range of speech.

CHAPTER XVIII.

The Motives of Speech--Expression--The Beginning of Human Speech--The Present Condition of Speech.

In vital economy, the search-light of science has found the protoplasm which from our present state of knowledge seems to be the first point of contact between elemental matter and the vital force. What secrets of biology remain unknown within the realm of life, only those who live in the future may ever know. In the first condition of vitalised matter we find the evidence of autonomy. Whatever may be the ultimate force which actuates this monad, the manifestations of its presence and the result of its energy are seen externally. Whatever may be the nature of that force which imparts motion to matter, the first impulse of the biod is to secure food or to associate itself with a unit of its own kind. This is perhaps the first act of volition within the sphere of life, the first expression of some internal want, and is the first faint suggestion of a consciousness, however feeble; and I may add with propriety, that it is my opinion that the vital and psychic forces operate in a manner not unlike the electric and chemical forces. They appear to polarise, and in this condition act on matter in harmony with that great law of Nature under which positive repels positive and attracts negative, and vice vers? We shall not attempt to follow the tedious steps of progress from inanimate matter to man, but begin with those intermediate forms which are so far developed as to utter sounds and understand the sounds of others. We will deal only with tangible facts as we find them. From whatever source expression may arise, or at whatever point it may appear, it is prompted by desire or some kindred emotion, either positive or negative.

[Sidenote: MODES OF EXPRESSION]

At the point where we begin to discuss this question there are two distinct modes of expression, either one of which can be used without the other. But I may mention here a cogent fact, that in the lower forms of life the normal mode of expression is by signs with supplemental sounds. In the higher forms, expression is by sounds, and signs are supplemental. And from the lower to the higher forms this transition is in harmony with the development of physical types. It occurs to me that signs were the first form of expression, and that sounds were first used to call attention to the sign made; and by an association of ideas the sounds became a factor of expression, and were used to emphasise signs. As we ascend the scale of life, sounds become more abundant, and signs less significant, and in the middle types they appear to be of nearly equal value, while in the higher tribes of man sounds are the normal mode of expression, and signs or gestures are used to emphasise them; and thus we see that signs and sounds in the development of the faculty of expression have quite changed places. This is consistent with the observed facts within the limits of human speech. There are tribes of mankind whose language is scarcely intelligible among themselves unless accompanied by signs; and it is said of some of the African tribes that their gestures are more eloquent than their speech. It appears to me consistent to believe that speech appears in the animal organism simultaneously with the vocal organs, and that the desire of expression must have preceded this. [Sidenote: PRESENT CONDITION OF SPEECH] The condition of the vocal organs depends upon the type of speech which they are used to utter, and the speech depends upon the quality of thought it is intended to express. That type of speech used by the Caucasian race within the space of a few centuries has developed from a vocabulary limited to a few thousand words into the polished languages of modern Europe, comprising new types and tens of thousands of new words, until to-day our own language contains more than two hundred and twenty thousand words, very few of which, however, if any, are entirely new. The phonetic elements on which is built up this huge vocabulary do not very greatly exceed in number those found in the lowest types of human speech in the world. The total number of these sounds does not much exceed two score in the highest forms of human speech; and about half this number can be shown as the vocal products of some species of the lower animals. Some philologists claim that the blending of consonant and vowel sounds is the mark which distinguishes human speech from the sounds uttered by the lower animals. To show how poorly this gigantic superstructure of fossilised science is supported by the facts, I have developed such effects in the phonograph from a basis of sounds purely mechanical, and without the aid of any part of the vocal apparatus of man or animal. The sounds from which I have developed such results were neither vowel nor consonant as those sounds are defined, but simply prolonged musical notes. In another chapter will be found some of the experiments which I have performed with the phonograph in the investigation of sounds of various kinds. If I am allowed to think for myself at all, I am not ready to accept as final some of the dogmas on the theory of sound which have long been held and taught, and many of which remain orthodox for no other reason than that no one has denied them. I am not ready at this point to spring upon the world any new theory of sound, but I am quite ready to refuse to believe some of the tenets set forth in the creeds of philology.

Heresy is the author of progress, and I confess myself a heretic on many of the current doctrines of the science of sounds.

CHAPTER XIX.

Language embraces Speech--Speech, Words, Grammar, and Rhetoric.

A definition of the word speech as used in this particular work is given elsewhere, and by this definition the word is used only in that sense which limits it to the sphere of oral sounds. It is that form of language which addresses itself only to the ear. The sounds which constitute it may be supplemented by signs or gestures, but such signs are only adjuncts, and are not to be regarded as an integral part of speech in its true sense. Speech cannot be acquired by those forms of life which occupy the lowest horizons of the animal kingdom, and have no organs with which to produce sound. In the light of modern use and acceptation language, broadly interpreted, includes all modes and means of communication between mind and mind. It therefore includes speech as one form, while signs or gestures constitute another form. Writing in all its various modes is another form of language. It may be substituted for either speech or gestures, but it does not thereby become speech in a literal sense, but within itself it constitutes another form of language. There seems to be some vague and subtle method of communication found in certain spheres of life which is called telepathy. While it is a mere ghost of language, so to speak, it has an identity which cannot be denied. This may perhaps be called another form of language.

By some eminent men of letters it is claimed that speech was invented, and therefore cannot be universally the same; and this is proven by the fact that different tribes of men have different tongues. They do not appear to realise, that to the first cardinal sounds of speech so much has been added age by age, by slow accretions, that the radex of speech is but a mere drop in the great ocean of sounds. The mobility of speech is such as to make it more susceptible to change than matter is; and yet we find that, by the laws of change, man has been evolved from a less complex state of matter, and that in these latter years he can only be identified as the descendant of his prototype by the most scrutinising care, and by picking up the dropped stitches in the great fabric of Nature. To illustrate the slow and imperceptible, yet never ceasing, never failing process of evolution, we may imagine a man picking up a single grain of sand at a certain point and carrying it a distance of a thousand feet, where he deposits it at another certain point; returning, takes a second grain of sand from the same place as he secured the first, and carries it to the point at which he deposited the first, and thus continues through his life. At his death his son succeeds him in the task, and continues through his life, and at the death of this man his son succeeds; and thus in turn each one succeeds the other through a million generations. Supposing the wind and rain left these grains of sand unmolested during this long lapse of time, it is evident that at the place from which the sand was taken there would be a hole, and where it was deposited there would be a hill. It is by such slight changes that Nature does her work; and thus it is that speech, as well as matter, has been transformed from what it was to what it is. The physical basis of life retains its identity through all those varied forms, from protozoa to the highest type; and so the phonetic basis of speech adheres through all the changing modes of thought and expression. Speech is the highest type of language and the most accurate mode of expression, and belongs only to the higher forms of the animal kingdom. It has passed through all inferior horizons coinciding with the mental, moral, and social planes through which man has passed in the course of his evolution.

[Sidenote: SPEECH AND WORDS]

Words are the factors of speech and the highest development of that faculty.

A word may be composed of one or more sounds so articulated as to preclude any interval of time between the utterance of any two of them, as "tune," in which the sounds appear to overlap and blend into each other. A single word may signify more than a single thing, and sometimes will suggest to the mind a category or group of connected thoughts, as "eat" or "telegraph," and such is the value of many of our words. This is especially true of words which combine two roots; but such a combination is usually found only in the higher types of human speech. But in these higher types words bear such relations to each other that we cannot well convey a complete idea with a single word; and hence it is that in the modes of expression used by man, each separate statement consists of two or more words bearing certain relations to each other, and these are often qualified by other words of less importance. This redundancy is due to the higher and more complex modes of thought used by man; and it is on such a state of facts that we have founded that branch of science called grammar, which would be of little use among those forms which occupy the planes of life inferior to man, and it is found of little use among the lower tribes of man, where it does not exist in any written form. Grammar does not make language, but serves as a kind of anchor by which the dialects of human speech are somewhat unified and made more stable; and to this is due in some measure the fact that savage tongues and dialects are more susceptible to change in their structure, while the phonetic basis upon which they rest remains the same.

[Sidenote: GRAMMAR AND RHETORIC]

In the more refined tongues of human speech, we go beyond that code of laws called grammar and amplify them into rhetoric. This branch of the science of speech could find no place among the lower types, as the words are few from which they may select; and so exact and arbitrary is the meaning of each one, and so uniform the relations, that no great variety of expression can be made with such a limited vocabulary. Their eloquence is in their brevity of speech. But while the types of speech used by the lower primates occupy a plane so low in the scale, they are as truly speech as the vocal organs that produce the sounds are truly vocal organs. Life is life, in what form soever it is found. It is not less real in the mollusc than in the man. The same is true of emotion, of thought, of expression, and of speech. Life, emotion, thought, expression, and speech began in embryo, and have developed co-ordinately with all the faculties possessed by man. They are as dependent upon each other as matter is on force, and as inseparable as light from energy. Speech is the physical manifestation of which thought is the ultimate force; it is a spoke in the chariot-wheels of consciousness; it is the body of which thought is the soul.

CHAPTER XX.

Life and Consciousness--Consciousness and Emotion--Emotion and Thought--Thought and Expression--Expression and Speech--The Vocal Organs and Sound--Speech in City and Country--Music, Passions, and Taste--Life and Reason.

At the beginning of life there is a consciousness which is not more feeble than is the life with which it is associated; and as that spark of life kindles into a flame, so that spark of consciousness kindles into the "ego," and nowhere can a line be drawn at which it may be said "here consciousness first intercepted life." But as the living form develops organs and members, through the agency of the vital force, whatever that may be, so consciousness develops into desires, emotion and thought. Where shall the line be drawn which separates these attributes? Standing in the centre, we look around and see the horizon touching the plain on every side, and this appears to us as a great circle, the centre of which is always occupied by the observer, and from our standpoint we imagine that everything between us and that horizon must be that distance from the centre; but as we move our point of view from place to place, we move the circle with us, and yet we cannot find the boundary line which marks this circle at any time. In a manner not unlike this we pass from centre to centre of the circles of life, and carry with us the circle, so that at no one point do we ever appear to be much closer to the horizon than we were at any other point.

[Sidenote: LIFE AND CONSCIOUSNESS]

The classification of genera and species is in a great degree arbitrary; but much less so than are these abstract characters of life and mind. There is nowhere a line at which emotion stops and thought begins; there is nowhere a line at which thought stops and expression begins; there is nowhere a line at which expression stops and speech begins. These blend into each other so that only by comparing the extremes can we discern a difference.

The tenets of metaphysics have heretofore been made to harmonise with the tenets of theology, and hence it is that we have learned to follow the laws laid down by others and not to think for ourselves. It has been as much a heresy to gainsay the dogmas of science as those of religion until recently; and even now the tender-footed doctors guard their theories with a vigilance and jealousy worthy of the angel that guarded the gates of Eden.

[Sidenote: CONSCIOUSNESS AND EMOTION]

Why should it be thought strange that monkeys talk? They see, hear, love, hate, think, and act by the same means and to the same end as man does. They experience pain and pleasure, to express which they cry and laugh just as man does. If the voluntary sounds they make do not mean something, why may those creatures not as well be dumb? If they do mean something, why may we not determine what that meaning is? It is true that their language is quite meagre and suited only to a low plane of life, but it may be the cytula from which all human speech proceeds, or it may be the inferior fruit borne upon the same great tree of speech. The organs of sensation in these creatures are modelled by the same design as those of man, are adapted to the same uses, and discharge the same functions. Then why should the vocal powers alone be abnormal, except in a degree measured by the difference of place which they occupy in the scale of Nature?

Social intercourse among men has been the chief means of developing human speech, and we find a true index to its condition in the social status of the different races of mankind; and by coming closer home, we find that even in different communities of the same race and within the limits of the same nation, a difference in the accuracy and volume of speech, which is measured by the difference of social culture. We find in rural districts, sparsely peopled and remote from the great centres of population, that speech is less polished and the number of words used greatly reduced in comparison to the same language used in the great cities and more populous communities, where, by reason of contact with each other and the constant use of speech, the vocal powers are much more developed and the command of language very much improved. This same law of development, inversely applied, would lead us in a direct line down through Nature, rank by rank, and we would find it a reliable unit of measure throughout the whole perspective of development.

The faculties of music, taste, and reason are measured by a like unit. It is difficult to trace the musical powers of animals, since music does not contribute to the comfort or development of types and only affords pleasure to the intellectual being, and hence is only an accomplishment obeying no rule of normal growth.

[Sidenote: THE FACULTY OF REASON]

As the use of the natural sense of taste makes possible the choice of nourishment, and all forms of life are thus sustained, the natural taste becomes an important factor of their comfort, and upon this physical basis rests, perhaps, the whole superstructure of ethics. The first idea of ownership is doubtless found in the possession of food; and this right of property is protected by the unwritten laws of incipient life. The faculty of reason, which man has arrogated to himself, is only limited by that dim line which bounds the vital sphere and sheds its rays through all the kingdom of life, from that point where the vital spark first lights the monad, through all the labyrinths of change, to man in the full pride of his divinity, standing upon the threshold of the angelic state. It is not by the exercise of reason that water flows down hill, or that matter obeys the law of gravity; but in the exercise of autonomy, however feeble may be the motive, reason guides the act. The power of this faculty is measured by the development of others, and there is no point between the two extremes at which reason intercepts life. The degree in which all the powers of sense and faculty are developed determines the horizon of the thing which possesses them. The aggregation of powers to act constitutes life; and the aggregation of powers to guide the action constitutes reason.

[Sidenote: ALL MAMMALS REASON]

Leaving the realm of metaphysics and returning to the order of primates, to which we shall confine our present work, I shall resume by repeating that not only do primates have the faculty of speech, but the whole family of mammals have some form of speech which is in keeping with their conditions of life. In addition to this declaration, I assert that all mammals reason by the same means and to the same ends, but not to the same degree. The reason which controls the conduct of a man is just the same in kind as that which prompts the ape. The latter cannot carry the process to such a great extent, but microsophic pedants have not shown in what respect the methods differ only in degree. That same faculty which guided man to tame the winds of commerce, taught the nautilus to lift its tentacles and embrace the passing breeze. Yet we are told that reason guides the man and instinct guides the nautilus. These are but two names for light; the one is dawn, the other noon, but both are light. I cannot see in what respect the light of a lamp differs from that of a bonfire except in volume; they are the products of the same forces in Nature, acting through the same media, and, becoming causes, produce the same effects. That psychic spark which dimly glows in the animal bursts into a blaze of effulgence in man. The one differs from the other just as a single ray of sunlight differs from the glaring light of noon. [Sidenote: EFFECTS OF ONE GREAT CAUSE] If man could disabuse his mind of that contempt for things below his plane of life, and hush the siren voice of self-conceit, his better senses might be touched by the eloquence of truth. But while the vassals of his empty pride control his mind, the plainest facts appeal to him in vain, and all the cogency of proof is lost. He is unwilling to forego that vain belief that he is Nature's idol, and that he is a duplicate of Deity. Held in check by the strong reins of theology and tradition, he has not dared to controvert those dogmas which bear the stamp of error on their face; he dares not turn away from the idols of his own conceit and read the rubrics written in the fossil rocks; he dares not take those proofs which none can counterfeit, and whose authority is not gainsaid; he dares not lay aside the yoke which galls the neck of patience, or breathe the air unblest by some mysterious rite performed in fear. By such restraints his ears are closed against those voices which appeal to him from without the temple gates of his belief. In what respect would man be less god-like if it be shown that monkeys talk? To elevate the humbler ranks could not degrade mankind. Whether man is the work of Deity or was evolved by laws of change from primal matter; whether he was made in one specific act or is the last amendment to a million prior types; whether he is the creature of design or accident, the authorship of his being and that of all the forms which roam the broad empire of life must be the same. We are all the effects of one Great Cause, whatever that may be, and that which gave to man the power of speech imparted it to apes; and I can see no reason why Nature should have drawn a line about this faculty, and made the rest a common heritage.

CHAPTER XXI.

Certain Marks which Characterise the Sounds of Monkeys as Speech-- Sounds accompanied by Gestures--Certain Acts follow certain Sounds--They acquire new Sounds--Their Speech addressed to certain Individuals-- Deliberation and Premeditation--They remember and anticipate Results-- Thought and Reason.

As a result of my experience with monkeys, I shall here sum up the chief points in which their speech is found to coincide with that of man, and note those features which distinctly characterise the sounds as a form of speech.

[Sidenote: SOUNDS OF MONKEYS AS SPEECH]

The sounds which monkeys make are voluntary, deliberate, and articulate. They are always addressed to some certain individual with the evident purpose of having them understood. The monkey indicates by his own acts and the manner of delivery that he is conscious of the meaning which he desires to convey through the medium of the sounds. They wait for and expect an answer, and if they do not receive one they frequently repeat the sounds. They usually look at the person addressed, and do not utter these sounds when alone or as a mere pastime, but only at such times as some one is present to hear them, either some person or another monkey. They understand the sounds made by other monkeys of their own kind, and usually respond to them with a like sound. They understand these sounds when imitated by a human being, by a whistle, a phonograph, or other mechanical devices, and this indicates that they are guided by the sounds alone, and not by any signs, gestures, or psychic influence. The same sound is interpreted to mean the same thing, and obeyed in the same manner by different monkeys of the same species. Different sounds are accompanied by different gestures, and produce different results under the same conditions. They make their sounds with the vocal organs, and modulate them with the teeth, tongue and lips, in the same manner that man controls his vocal sounds. The fundamental sounds appear to be pure vowels, but faint traces of consonants are found in many words, especially those of low pitch; and since I have been able to develop certain consonant sounds from a vowel basis, the conclusion forces itself upon me that the consonant elements of human speech are developed from a vowel basis. This opinion is further confirmed by the fact that the sounds produced by the types of the animal kingdom lower than the monkey, appear to be more like the sounds of pipe instruments; and as we rise in the scale, the vocal organs appear to become somewhat more complex, and capable of varying these sounds so as to give the effect of consonants, which very much extends the vocal scope. The present state of the speech of monkeys appears to have been reached by development from a lower form. [Sidenote: EACH RACE HAS ITS PECULIAR TONGUE] Each race or kind of monkey has its own peculiar tongue, slightly shaded into dialects, and the radical sounds do not appear to have the same meaning in different tongues. The phonetic character of their speech is equally as high as that of children in a like state of mental development; and seems to obey the same laws of phonetic growth, change, and decay as human speech. It appears to me that their speech is capable of communicating the ideas that they are capable of conceiving, and, measured by their mental, moral, and social status, is as well developed as the speech of man, measured by the same units. Strange monkeys of the same species seem to understand each other at sight, whereas two monkeys of different species do not understand each other until they have been together for some time. Each one learns to understand the speech of the other; but, as a rule, he does not try to speak it. When he deigns an answer, it is usually in his own tongue. The more fixed and pronounced the social and gregarious instincts are in any species, the higher the type of its speech. They often utter certain sounds under certain conditions in a whisper, which indicates they are conscious of the effect which will result from the use of speech. Monkeys reason from cause to effect, communicate to others the conclusion deduced therefrom, and act in accordance therewith. If their sounds convey a fixed idea on a given subject from one mind to another, what more does human speech accomplish? If one sound communicates that idea clearly, what more could volumes do? If their sounds discharge all the functions of speech, in what respect are they not speech?

[Sidenote: CANNOT THINK WITHOUT WORDS]

It is as reasonable to attribute meaning to their sounds as to attribute motives to their actions; and the fact that they ascribe a meaning to the sounds of human speech, would show that they are aware that ideas can be conveyed by sounds. If they can interpret certain sounds of human speech, they can ascribe a meaning to their own. They think, and speech is but the natural exponent of thought; it is the audible expression of thought, and signs are the visible expression of the same; born of the same cause, acts to the same end, and discharges the same functions in the economy of life. To reason is to think methodically; and if it be true that man cannot think without words, the same must be true of monkeys. I do not mean, however, to claim that such is a fact with regard to man thinking; but if such can be shown to be a fact, it will decide the question as to the invention of human speech, as it was necessary for man to think in order to invent; and, by the same rule, he could not think a word which did not exist, and therefore could not have invented it. But I beg to be allowed to stand aside and let Prof. Max M 黙 ler and Prof. Whitney, the great giants of comparative philology, settle this question between themselves; and I shall abide by the verdict which may be finally reached.

But theories are useless things when the facts are known; and since I have actually learned from a monkey a certain sound having a certain value and meaning a certain thing, and by repeating that sound to a monkey of the same species have met with uniform results, have understood him, and been understood by him, no argument could be so potent as to cause me to believe that this was accident. I am aware that coincidents occur; but when they become the rule instead of the exception, they are no longer mere coincidents, but are the normal state of things.

[Sidenote: THOUGHT AND REASON]

In conclusion, I would say that since the sounds uttered by monkeys perform all that speech performs, is made of the same material, produced by the same means, acts to the same ends, and through the same media, it is as near an approach to speech as the mental operations by which it is produced are an approach to thought. If it can be shown that these mental feats are not thought, the same process of reasoning could show that these sounds are not speech. If man derived his other faculties from such an ancestry, may not his speech have been acquired from such a source? If the prototype of man has survived through all the vicissitudes of time, may not his speech likewise have survived? If the races of mankind are the progeny of the Simian stock, may not their languages be the progeny of the Simian tongue?

CHAPTER XXII.

The Phonograph as an aid to Science--Vowels the basis of Phonation--

Consonants developed from a Vowel basis--Vowels are Compound--The Analysis of Vowels by the Phonograph--Current Theories of Sound--Augmentation of Sounds--Sound Waves and Sound Units--Consonants among the Lower Races.

The application of the phonograph to my special work is really the discovery of a new field of usefulness for that wonderful instrument, which, up to this time, has held the place of a toy more than that of a scientific apparatus of the very highest importance in the study of acoustics and philology. In many ways the use of this machine is so hampered by the avarice of men as to lessen its value as an aid to scientific research, and the Letters Patent under which it is protected preclude all competition and prevent improvements. However, I have been able, even with the poor machines in general use, to discover some of the most important facts upon which are based the laws of phonation. I shall here attempt to give in detail but a few of these experiments, as they are yet crude, and in some cases the deductions therefrom not positively certain. [Sidenote: VOICES OF MEN AND MONKEYS] From the various records that I have made of the voices of men and monkeys, I am prepared to say that the difference is not so great as is commonly supposed, and that I have converted each one into the other. I would not be understood to say that I have done this with all their sounds, nor that the monkey's sounds were converted into human speech, but the fundamental sounds of each were changed into those of the other. I find that human laughter coincides in nearly every point with that of monkeys. They differ in volume and pitch. By the aid of the phonograph I have been able to analyse the vowel sounds of human speech, which I find to be compound, and some of them contain as many as three distinct syllables of unlike sounds. From the vowel basis I have succeeded in developing certain consonant elements, both initial and final, from which I have deduced the belief that the most complex sounds of consonants are developed from the simple vowel basis, somewhat like chemical compounds result from the union of simple elements. Without describing in detail the results, I shall mention some simple experiments which have given me some very strange phenomena. I dictate to the phonograph a vowel in different keys while the cylinder rotates at a given rate of speed. I then adjust the speed to a certain higher or lower rate and follow the results. By reversing the motion of the cylinder the sounds are reduced to their fundamental state. By this means we eliminate all familiar intonation, and disassociate it from any meaning which will sway the mind, and in this way it can be studied to advantage. [Sidenote: THE SOUND WAVES] At a given rate of speed I have taken the record of certain sounds made by a monkey, and by reducing the rate of speed from two hundred revolutions per minute to forty, it can be seen that I increased the intervals between what is called the sound waves and magnified the wave itself fivefold, at the same time reducing the pitch in like degree, and by this means I could detect the slightest shades of modulation. I may remind you here that in this process all parts of the sound are magnified alike in all directions, so that instead of obtaining five times the length, as it were, of the sound unit or interval, we obtain the cube of five times the normal length of every unit of the sound. The slightest variation of tension in the vocal chords may be detected, and every part of the sound compared to every other part.

Having thus augmented the quantity of sound, by increasing alike the sound unit and interval, it can be recorded on another cylinder and multiplied again as long as the vibrations can produce sound. From the constant relation of parts and their uniform augmentations under this treatment, it has suggested to my mind the idea that all sounds have definite geometrical outlines, and as we change the magnitude without changing the form of the sound, I shall describe this constancy of form by the term contour.

In a few instances I have been able, by reducing the record of certain sounds from a high pitch to a lower one, to imitate the sound thus reduced with my own vocal organs, then by restoring this record of my voice to its normal speed have obtained almost a perfect imitation of the sound. This effect, however, does not always follow, and in many instances my best imitations have not developed the original at all. But this presents a new problem in acoustics. I must here take occasion to say that the difference of pitch, quality, &c., in sounds does not appear to me to depend alone upon the length of the sound unit, but there seems to be a difference of ultimate form and mode of propagation which have much to do with the contour of the developed sound.

[Sidenote: THE SOUND FORCE]

By mode of propagation I mean the organs brought into use for the purpose of producing the sound, the apertures through which the sound force passes, and the auxiliaries by which it is moulded into certain shapes. By ultimate form I mean the geometrical shape of the sound force when first converted into sound. That there is such a thing as form has been clearly demonstrated by the phoneidoscope. Prof. John B. De Mott has very kindly aided me in reducing certain sounds to a visible condition. I had conceived an idea before this that if the path described by the energy which produced sound could be made visible, that it would be found to have the form of a convolute spiral, that these spirals recede from the centre or point of propagation in every direction like the radii of a sphere, and that that aspect of sound which we call waves, is simply the point at which these spirals intercept each other, which of necessity would be of uniform distance from the centre, increasing at each successive point throughout the entire sound-sphere or space through which the sound passes in all directions from the centre to infinity. I shall refrain from discussing this point till such a time as I can show at greater length my reasons for this belief. I may add here that I have made records of the human voice with which I have deceived the monkeys, and I have made records of the monkey's voice with which I have deceived the very elect of linguists and musicians. Some critic once remarked to me that the sound made by a monkey was not really laughter, but only a kind of good-natured growling. This may be correct, but the same is true of human laughter, as the one may be converted into the other, and a good-natured growl expresses the emotion which is felt by man as well as monkey.

The phonograph shows that they are identical in sound and form, besides the fact that they are the outburst of the same passion, actuated by the same cause and executed by the same muscles, so that their identity, mentally, physically, and mechanically, is the same.

[Sidenote: VOWEL SOUNDS]

Among the sounds of the Simian voice I have not found the English vowels "a," "i," or "o," except, perhaps, "i" short as sounded in the word "it." The vowel "u," as sounded like "oo" in "shoot," seems to be the chief sound of their speech. One important point which I discovered from the phonograph is, that sounds or tones which are purely musical are reproduced alike with the cylinder turning either way, while all speech sounds are slightly changed when the cylinder is reversed, which shows the sounds to be compound. I find that "w" may be developed from any consonant by manipulating the cylinder of the phonograph, and it is a fact also that the initial consonant imparted to any vowel does not continue through the vowel. This I have shown by making a vowel sound which I prolong for some seconds with the cylinder revolving at a given rate of speed. While reproducing this at a normal speed I intercepted at any point, and developed the sound "w" as heard in "woe." The instant I have blended this into the vowel, I lift the diaphragm until the normal speed is restored, when I replace the reproducing tooth showing the sound without the consonant. In like manner I dictate to the phonograph any vowel sound preceded by a consonant. The consonant I utter in a natural way, the vowel I prolong for some seconds, and in the act of reproducing this I cut the sound in two and find the vowel element is not modified by the consonant which preceded it, hence, I observe that the consonant merely suggests to the mind a certain form of sound which does not change the fundamental vowel. In fact, it aids the voice somewhat in uttering the vowel.

If human speech were composed of none but vowel sounds the human voice could scarcely utter them in a continued conversation; their monotony would not so much offend the ear as it would try the vocal powers, and man would soon acquire consonants to aid the voice if for no other use.

[Sidenote: DOUBLE AND TREBLE CONSONANTS]

Among the Simians the better types of speech show this tendency, and in the lower types of human speech we find all the vowel elements, while consonants are not by any means so numerous. Another fact is this, among the lower races of mankind double consonants are rare, and treble more so. Of course their tongues consist of fewer words, as has been shown before, which paucity arises from their few wants and simple modes of life, and hence the scope of vocal growth is much contracted. Beginning with the lowest tribes of men, we find the consonants increase in number and complexity as we ascend the scale of speech. To this, perhaps, is due the fact that the Negroes now found in the United States after a sojourn of two hundred years with the white race on this continent are unable to utter the sounds of "th" "thr," and other double consonants. The former of these they pronounce "d" if breathing, and "t" if aspirate. The latter they pronounce like "trw" or "tww." The sound of "v" they usually pronounce "b," while "r" resembles "w" or "rw" when initial, but as a final sound is usually suppressed. They have a marked tendency to omit auxiliary and final sounds, and in all departures from the higher types of speech tend back to ancestral forms.

I believe if we could apply the rule of perspectives and throw our vanishing point far back beyond the chasm that separates man from his Simian prototype, that we would find one unbroken outline tangent to every circle of life from man to protozoa in language, mind, and matter.

CHAPTER XXIII.

The Human Voice--Human Bagpipe--Human Piccolo, Flute, and Fife--The Voice as a Whistle--Music and Noise--Dr. Bell and his "Visible Speech."

One of the very curious feats which I have performed with the phonograph is the conversion of the human voice into the sounds of various instruments. I had my wife sing the familiar Scotch ballad, "Comin' through the Rye," to the phonograph while the cylinder was rotating at the rate of about forty revolutions per minute. Each word in the song was distinctly pronounced and the music rendered in a plain, smooth tone. I then increased the speed of the machine to about one hundred and twenty per minute, at which rate I reproduced the song. It was a very perfect imitation of the bagpipe with no sign whatever of articulation. The melody was preserved with only a change of time. The speech character was so completely destroyed that I repeated this record to a large audience in which were several eminent musicians, not one of whom suspected that it was not a real bagpipe solo. In like manner I have converted the sounds of the voice into a very perfect piccolo, flute, fife, and into a fairly good imitation of a whistle sound. To produce the whistling effect and the fife sound the rate of speed must be necessarily very high, and some notes will not be perfectly converted for some reason which I have not yet fully understood. Some voices are much more easily converted into the flute effect than others. To get the best flute sounds, a full, smooth, mezzo-soprano gives the best effect. In reversing the operation, the sounds of these instruments can be made to imitate the human voice somewhat, but not exactly, not only in the fact that the modulation is wanting and there is no semblance to consonant sounds, but the tone itself differs in quality from that of the voice.

[Sidenote: CONTOUR OF SOUNDS]

Among other respects in which the vocal sounds of man and Simian resemble is in the contour of the sounds, which I have defined elsewhere. I have called attention to the fact that by reversing the cylinder of the phonograph and repeating the sound recorded thereon that a musical note or sound would repeat alike each way. Most of the sounds made by other animals do this, but those made by man and Simian alike show modulation, not, however, equally distinct. The notes of birds repeat alike both ways except their order is reversed. Again, to magnify the sounds as I have shown it can be done, allows you to inspect them, as it were, under the microscope, and this examination shows the contour of the sounds of these two genera to resemble.

Dr. Alexander Melville Bell has shown, in his work on "Visible Speech," that the organs brought into use in the production and modification of sounds must work in harmony with each other; hence it is that by a study of the external forms of the mouth the movements of all the organs used in making any sound can be determined with such certainty that deaf-mutes can be, and have been, successfully taught to distinguish these sounds by the eye alone. And it was by such a method that I set out to read the temple inscriptions from the ruins of Palenque, some years ago, at which time I had not heard of Dr. Bell's learned and excellent work. The main feature of those glyphs, by which I was guided, was the outline of the mouth, which the artist had sought to preserve and emphasise at the cost of every other feature, and by this process I found to my satisfaction some ten or twelve sounds or phonetic elements of the speech used by these people; but not knowing the meaning of the sounds in that lost tongue, I did not attempt to verify them, but when I find the time to devote to them I believe I can accomplish that.

[Sidenote: TRIP TO AFRICA]

It is a part of my purpose, in my trip to Africa, to try to secure photographs of the mouths of the great apes while they are in the act of talking, and to this end I am having constructed an electric trigger, with which to operate my photo-camera at long range, and I shall try to furnish to the eminent author of "Visible Speech" some new and novel subjects for study.

CHAPTER XXIV.

Some Curious Facts in Vocal Growth--Children and Consonants--Single,

Double, and Treble Consonants--Sounds of Birds--Fishes and their Language--Insects and their Language.

[Sidenote: SOUNDS UTTERED BY CHILDREN]

I shall take occasion here to mention some curious experiments, which have suggested themselves to me in my work with the phonograph. For lack of time and opportunity, I have not carried them far enough to give exact and final results; but it has occurred to me that philology may be aided by taking a record of the sounds made by a number of children daily through a period of two or three years from birth. The few experiments which I have tried in this particular line are sufficient to show that the growth of speech obeys certain laws in the development of vocal power. It is apparent to me that the first sounds uttered by children have no consonants, and that certain consonants always appear in a regular succession and always single. The double consonants develop later, and the triple consonants appear to be the last acquirement. I have not the space to go to great length on this subject, and my experiments have not been sufficient to enable me to formulate with certainty any set of rules by which the development of this faculty is uniformly governed.

It is my purpose, on my return from Africa, to set on foot a series of such experiments, with the hope of ascertaining the facts connected therewith. And while in Africa I shall aim to make such records of the natives as to ascertain whether their speech conforms to the same laws of development or not. It is my earnest hope to be able to do the same thing with the great apes which I am going chiefly to study. I think if I can record on a phonograph cylinder the sounds uttered by a young chimpanzee under certain conditions once each day for a year or so, I can determine whether there is a like growth in their speech, and to what extent the same laws control it. I have already observed that the quality of voice in a given species of monkey changes with his age very much in the same manner as the human voice; but I have not been able to follow the changes through one individual specimen by which to ascertain the exact manner of such change.

[Sidenote: SOUNDS OF BIRDS]

The sounds of birds have been studied perhaps more than any others except those of man, but they have not been studied as speech, nor to ascertain their meanings. Their musical character has attracted attention and been the subject of some discussion. My opinion is that much that has been said on that subject belongs more properly to the realm of poetry than of science. I think the sounds of birds are chiefly intended for speech, but it may supply the place of music in their 鉎 thetic being; but, so far as I have observed, I confess that I cannot find that they obey the laws of harmony, melody, or time, and it is my opinion that most of the efforts to write the sounds of birds on a musical staff are not to be relied upon as accurate records of the sounds. There is no doubt that each sound uttered by a bird is in unison with some note in the chromatic scale of music, but the intervals between the tones of the same bird do not coincide with those of the human voice. It is quite evident that birds possess an acute sense and ready faculty for music, and many of them show great aptitude in imitating the sounds of musical instruments; some varieties of birds, such as the southern mocking-bird, the thrush, and others, imitate with great success the sounds of other birds. They often do this so perfectly as to deceive the species to which the sounds belong. The songs of birds, as they are called, appear to afford them great pleasure, and they often indulge in them, I think, as a pastime; the effect is pleasing to the ear because of its cheerfulness, and it is not discordant or wanting in richness of tone in most birds. From the little study I have given them I think it safe to say that the range of sounds possessed by any one bird is quite limited and their notes are strictly monophones. This last remark does not apply to the sounds made by parrots and birds of that kind.

The parrot is perhaps possessed of the greatest vocal power of any other bird. He imitates almost the entire range of sounds that are uttered by all other birds combined, and can also imitate the sounds of human speech from the highest to the lowest pitch of the human voice. In addition to all this, he imitates many noises, such as the sounds of sawing wood, the slam of a door, and the whistling of the wind. The vocal range of the parrot is perhaps the most marvellous of all the vocal products of the animal kingdom. One strange thing, however, that I observe among them is, that the range of sounds that they use among themselves is very small. I have made some records of parrots, macaws, cockatoos, &c., and I find their natural vocal sounds usually wanting in quality: most of their sounds are hoarse and guttural.

Among the gallinaceous birds there does not appear to be much music.

There is a great sameness of sounds in the different species, and they seem to be confined to the economic use of speech.

In my early life I devoted much time to gunning, and I observed then, and called attention to the fact, that when a covey of birds became scattered I could tell at what point they would huddle. I could tell this by the call of one bird and the reply of the others. The call-bird, which was always the leader of the covey, would sound his call from a certain point near which the other birds would usually assemble, and during this time they would answer him from various other points. The sound used by the call-bird is unlike that used by the rest of the flock, but the sounds with which they reply to him are all alike, and by observing this I could always find the covey again by allowing them time to come together, especially if it was late in the afternoon.

Mr. Wood, of Washington, D.C., has given such attention to the sounds of birds that he can interpret and imitate nearly all the sounds made by domestic birds, and many of those made by wild birds. He has twice confused and arrested the flight of an army of crows by imitating the calls of their leader. His feats have been witnessed with astonishment by many men of science.

[Sidenote: SOUNDS OF FISHES]

Among fishes I have found but few sounds, and most of these I have never heard except when the fish was taken out of the water. The carp and high-fin, however, I have frequently heard while in the water. It has occurred to me that the sound is not the medium of communication, but it is the result of an action by which they do communicate even when the sound is not audible. I have observed while holding the fish in my hand when he makes this sound that it produces a jarring sensation which is very perceptible. It is quite possible that in his natural element these powerful vibrations are imparted to the surrounding water, and through it communicated to another fish, who feels it in his sensitive body instead of hearing it as sound. It may be accompanied by the sound merely resulting from the force applied, but not in itself constituting any part of the means of communication. It is not unlike what we call sound, in the fact that it is generated in the same way, transmitted in the same way, and received in the same way as sound. When I have time and opportunity I shall carry my studies of the language of fishes much farther. Their means of communication are very contracted, but it is superfluous for me to say that they have such means.

Many observations have already been made on the language of insects, and much diversity of opinion prevails. Very little has been said about the details of their intercourse, but the consensus of opinion is that they must in some way communicate among themselves. To me they seem to live within a world of their own, as other classes of the animal kingdom do. The means of communication used by mammals could not be available among aquatic forms, any more than could their modes of locomotion. Each different class of the animal kingdom is endowed with such characters and faculties as best adapt them to the sphere in which they live; and the mode of communication best fitted to the conditions of insect life would be as little suited to mammals, perhaps, as the feathers of a bird would be for locomotion in the realm of fishes.

[Sidenote: LANGUAGE OF INSECTS]

I am aware that some high authorities have claimed that insects communicate by sounds. My own opinion is that they employ a system of grating or scratching by means of their stigmata, but that the sound created thus performs no function in the act of communicating, but is only a bi-product, so to speak, and that the jarring sensation transmitted through the air is the real means by which they understand each other, possibly somewhat like telegraphy, in which the sounds are not modulated, but are distinguished by their duration and the interval between them. I do not announce this as conclusive, but merely suggest it as a possible key to their mode of intercourse.

[Sidenote: A COLONY OF ANTS]

I have observed that signs prevail to a great extent among ants. Some years ago I had an opportunity of studying a colony of ants, and I watched them almost daily for several weeks. I had seen it stated that they found their way by the sense of smell, but these observations confirmed my doubts on that point, and I feel justified in saying that they are guided almost, if not entirely, by landmarks. On the bark of a tree from which they were gathering in their winter stores, I observed that there were certain little knots or protuberances by which they directed their course and which they always passed in a certain order. Between these landmarks they did not confine themselves to any exact path, but the concourse would sometimes widen out over the space of more than an inch, but as they approached a landmark every ant fell into line and went in the exact path of the others, which rarely exceeded in any case more than an eighth of an inch in width. Whenever an ant would lose its way it would lift its head high into the air, look around, and then turn almost at right angles from the course it was pursuing towards the path of the others. In scores of cases I observed that the outward-bound ant, when it had been lost and returned to the path, always came on the homeward side of the landmark and passed out. On the other hand, if a homeward-bound ant was lost it would approach from the outward side of the landmark and pass in. About five feet from the ground were two small, round knots, about one-eighth of an inch in height, and a space between them of about the same width. This appeared to be one of their most conspicuous and reliable landmarks, and every ant that I saw pass in or out during the lapse of weeks passed between these two points. The burdened ant always appeared to have the right of way, and when meeting another without a burden there was no question of this right. In such a case the burden was usually held aloft, and the right of way conceded without debate. A little later in the season I had the opportunity of seeing the same colony emigrate to a point about eighty feet distant from their original abode, at which time they carried large burdens and were many days in completing their work, but the same system and methods prevailed.

As far as desire can be found in life the means of expression go hand in hand with it, but I do not contend that desire alone is the origin of this faculty. So far as human ears can ascertain, the lowest forms of life appeared to dwell in perpetual silence, but there may be voices yet unheard, more eloquent than we have ever dreamed of.

CHAPTER XXV.

Facts and Fancies of Speech--Language in the Vegetable Kingdom--Language in the Mineral Kingdom.

In the early part of this work I have recorded the material and tangible facts with which I have dealt, and have not departed from such facts to formulate a theory beyond a working hypothesis. I have not allowed myself to be transported into the realm of fancy, nor have I claimed for my work anything which lies beyond the bounds of proof. But in the wide range through which I have sought for the first hint of speech, it is only natural that many theories have suggested themselves to me from time to time, some of which would appear almost like the dreams of hasheesh. But while they are like the fairyland of speculation, they are not more wild and incoherent than are many of the dogmas of metaphysics. And at this point I shall digress from my text so far as to say that I have followed the motives of language through the higher planes of life and thence downward to the very sunrise to the vegetable kingdom, and on through the dim twilight across the mineral world to that point where elemental matter is first delivered from the hands of force. Standing upon the elevated plane of human development, it is difficult for man to stoop to the level of those inferior forms from which he is so far removed in all his faculties; but if his senses could be made so delicate as to discern the facts, he would find perhaps that in the polity of life all horizons are equidistant from each other. But looking back from where he stands, his powers fail to reach the real point of vital force at which all life began, and his contracted senses bring the vanishing point of this perspective far into the foreground of the facts.

From the highest type of human speech to the feeblest hint of expression there is a gradual descent, and at no point between these two extremes can there be drawn a line at which it may be said "here one begins, and here another ends." The same is true of other faculties; and from the vital centre at which matter first receives the touch of life to the circumference of the vital sphere, all powers radiate alike, and there is no point that I can find between that centre and infinity at which some new endowment intercepts the line.

Descending the scale of life by long strides, from man to the lowest form of zooids, we cannot designate the point at which a faculty is first imparted to the form which has it, and this truth extends throughout the vital cosmos.

[Sidenote: LANGUAGE IN THE VEGETABLE KINGDOM]

The line of demarcation which separates the animal and vegetable is but a wavering, blended mezzotint, and the highest forms of vegetable life seem to overlap the lowest forms of animal, so far that no dividing line is positively fixed. The highest types of vegetable seem to have the faculty of expression in a degree corresponding to, and in harmony with, the rest of their organism. I do not mean to say that the impulse under which a plant acts is synonymously with that which prompts the animal, but both appear to be the effect of the same cause.

In some forms of vegetation the selection of food of certain kinds and the aversion to other certain kinds, would indicate that the organism is capable of design and purpose in a degree perhaps much higher than some of the lowest forms of the animal kingdom. The reaching out of roots in search of food in the earth, the opening and closing of leaf and bloom, seeking the moisture and carbon from the atmosphere, suggest a feeble expression of desire. The choice of food is so well defined in some plants as to indicate a power of selection far greater than some protozoans exercise. It is a known fact that a change of food and conditions often modify a plant in such degree as to make it difficult to recognise except by the technical laws of classification, and yet its identity is not lost. Such changes do not effect all plants in the same degree, as some of them will undergo a change of diet or conditions without material effect. In many instances a marked dislike to certain kinds of food has been observed, and the sensitiveness of some plants is shown in the foliage, bloom, and even in the roots.

[Sidenote: LANGUAGE OF THE MINERAL KINGDOM]

In passing from the vegetable to the mineral kingdom, we find a like diffusion of types overlapping and blending into each other. Some forms of vegetation are so low in the scale of organism as to make it difficult to say whether they are vegetable or mineral compounds. Of course we find no trace of speech, but there is that hint of expression or suggestion of desire as found in the vegetable kingdom. In the chemical world one element will select another with which it will combine, while to other elements it shows a great aversion. When one chemical element selects another and combines with it we call this chemical affinity. The ultimate force which causes this affinity is one of the unknown facts concerning matter; but it is possible that such affinities and aversions constitute the basis upon which rests the selections and aversions of plants and animals. But as we rise in the scale the combinations of matter become more complex and the functions of each part more specific. It is possible, when we become more familiar with the forces of Nature, that we shall find that affinity and repulsion are but the positive and negative poles of the forces which act on matter; that chemical, vegetable and animal activity are based upon the same fundamental causes, and that speech, which is only one form of expression, is the highest product of such an ultimate force, but in all conditions of matter, such forces, either positive or negative, are the ultimate motives of expression.

[Sidenote: VITALISATION OF MATTER]

As chemical formulas differ from each other without losing the identity of their elements which constitute them, so animal organisms and plant forms differ as the spheres of life to which they are assigned differ. It is possible that chemical affinity may be the germ from which all language springs, as the chemical elements are the materials from which all compounds are built up. The vitalisation of matter itself, and the arrangement of the ultimate particles which constitute a living body, are the work of the vital force in a polarised condition. This will account, in a measure, for all the individuals of one type selecting one mode of expression, as they select or conform to one physical outline. In every rank of life there seems to be some intuitive mode of expression which suggest itself to all the individuals of that kind when they desire, under the same conditions, to express the same thing. The exceptions to this law of expression increase in number as we rise in the scale of life, and the means of expression increase and widen and the faculty of thought enlarges. The laws of chemical affinity are rigid and uncompromising, and there are but few exceptions in them, and only marked changes of condition can modify the results. As we ascend even in the mineral kingdom to the higher compounds we find a wider range of variation; and as we continue our ascent through the vegetable world, we find the same, and on through animals to the highest type. In the lower planes types are more strictly adhered to, habits and food more rigidly observed, while among the highest types of cultivated plants we find a great diversity of fruit and bloom, the capability of transplanting and the creation of new species, without losing the generic identity of the plant or even making it questionable. In the animal kingdom the same law is complied with; and step by step as we ascend the same types show greater and greater diversity, until we reach man--the climax of all life, and within his genus, variation knows no bound.

In conclusion, I may say that man as he now is has the faculty of speech. It is reasonable to believe that he has always had this faculty since he was man. If there has ever been a time in the history of his organism when he acquired his being from some progenitor which was not man, he acquired at the same time the faculty of speech, and that progenitor did not impart a thing which he did not have. While it is true that speech, as I have used it, is confined to vocal sounds, other modes of expression have preceded it, and such has been a common faculty inherent through all forms and planes of life. I am aware that two ingredients combined may make a compound unlike either one, and such may be the case with speech, but the elements which constitute the compound must have been for ever present.

CHAPTER XXVI.

THE SPEECH AND REASON OF DOMESTIC ANIMALS.

Dash and the Baby--Two Collies talk--Eunice understands her Mistress--Two Dogs and the Phonograph--A Canine Family--Cats and Dogs--Insects--Signs and Sounds.

[Sidenote: THE SPEECH OF DOMESTIC ANIMALS]

To those who are familiar with rural life, there can be nothing strange in hearing it said that all animals can talk among their kind. Among the daily incidents of farm life, there occur so many proofs of this as to place the question beyond debate. The cattle have means of conveying ideas to other cattle, and sheep and hogs understand other sheep and hogs, and the means employed are sounds. These sounds are used in the same way that man uses them to convey his thoughts, and since they discharge all the functions of speech, in what respect are they not speech? The types of speech differ in different genera, as their physical types do, but they are not any the less speech on that account. Among the domestic animals, I think the dog has, perhaps, the highest type of speech; and this is doubtless, in some measure, due to his intimate relations with man, from whom he has learned and added a little to his mental store, and this must find an outlet through speech. That dogs think and reason is not to be doubted by the most stupid observer, and they often make known their thoughts so that even man can interpret them with certainty; but the speech by which they express those thoughts is of course rudimentary. Dogs often discharge certain duties with such promptness that bigots declare that it is mechanical and done without motive, but there are many thousands of cases where the dog has assumed and performed duties of others, entirely outside of his own sphere, which nothing but reason could have prompted.

When I was only a few weeks old, my father had given to him a little white poodle, which he called Dash. He was about my own age, and we grew up together. In those days, children were rocked in the old-time cradle, and I, like other babies, had a cradle. When I was a few months old, on one occasion I was left asleep in my cradle, and no one was in the room but Dash and myself. Having been disturbed in my sleep, I woke up and cried, and Dash, seeing the condition of things, came to the cradle, and, rearing on his hind feet, rocked it with his paws, and whined and barked until I had gone to sleep again. My mother has often told me of this, and assured me that he had never been taught to do this, but always after practised it, not only with myself, but with my younger brothers and sisters, until, at the age of thirteen, he came to an untimely death at the hands of a bull-dog, whose name and tribe I have never ceased to hate. I gave Dash the burial that he deserved, and had a long procession of mourning children follow his remains to the grave, where I delivered the funeral sermon, and we all sung a hymn. About three years ago, in company with an older sister, I visited the spot for the first time in nearly thirty years, but no sign of the little grave remained.

What else but reason could have prompted this act? The dog had seen it done by human beings, and had noted the result. Whether his whining was intended as singing or not, I am unable to say, but from my recollection of seeing him do this with the younger children, I believe that it was intended to soothe or entertain, and his barking to call some one into the room.

A farmer by the name of Taylor, living in East Tennessee, some years ago owned two very fine collies, and they had been trained to drive the cattle and sheep about the farm, to drive strange cattle away from the premises, to guard the gates or gaps opened temporarily for hauling about the farm, and many similar duties. On one occasion, in haymaking time, as night was approaching, the waggon made its last homeward trip for the day, and the men working in the meadow prepared to go home. The driver of the waggon, supposing the men from the meadow were following and would close the gates, left them open, and one of these was between the corn-field and a pasture containing a number of cattle. The men, however, did not follow the waggon, but took a near way across the field, and the gate was left open. While the family was at supper, one of the collies was restless and barked continually, and gave such signs of uneasiness as to assure all that something was wrong. His master went to the door, and the dog ran to the gate in the front of the house, and continued barking and lashing his tail with great energy. The master followed to the front gate, and the dog immediately ran barking down the road, but looking back from time to time to see that his master followed, which he did, and was thus led to the open gate, where he found the other collie on guard and keeping the cattle from passing, which they were trying to do. What less than reason could have prompted these dogs to such an act? And what less than speech could have enabled them to execute this feat? They observed the neglect or error of the driver, and foresaw the evil consequences, and it could only have been by agreement reached through an interchange of thoughts that one of them watched while the other gave the alarm. I have known some of these dogs that knew certain cattle by name, and would go into the herd and drive out the one whose name was designated, while it is true in other cases that the dog would only drive out such as were pointed out to him. But many instances proved that they are able to learn the names of the cattle. It is certain that in many instances dogs know the names of the children belonging to the family, and often distinguish them by name. I presume no one doubts that they learn their own names, so that each dog may know when he is called. I know a dog, now living near Leominster, Mass., that extinguished an accidental fire which had been caused by the hired man carelessly dropping a burning match in some straw in the barn-yard after lighting his lantern. The dog had to fight the fire with his paws, and by the time he had extinguished it they were much singed. His loud barking was sufficient to warn the family that something unusual was taking place. They soon responded to his call, and found that he had the fire quite under control. He had thus saved his master's barn and house from the flames, and since that time, as I have witnessed myself, will not allow any one to light a cigar with a match in his presence. The peculiar sound which he makes under such circumstances appeals to the sense of fear or apprehension, and I have observed that the significance of all speech depends much upon intonation. It is less so with man, perhaps, than with other animals, because of the great number of words which amplify and shade his meanings. But by a single word of human speech we can express many shades of meaning simply by modulation; but having at our command so many words to qualify our meaning, we lose sight of the value and power of intonation. The difficulty of discerning the delicate shades of meaning imparted by intonation, depends upon the mode of thought, and the simpler this is the keener the power to interpret inflections. One very important fact is that a dog only learns to interpret one sound on one subject at any one time. He cannot put together in his mind a great number of sounds, nor interpret complex ideas in detail. I know a dog in Charleston, South Carolina, that would fly into a rage and bark fiercely if you say, "Chad, where is that big black dog that whipped you so badly?" But repeated experiments proved to my mind that the dog did not interpret any part of the sentence except the words "black dog," and even this seemed to depend chiefly upon the sound "black," and by saying this sound you would get the same results as to use the entire sentence. He had been whipped by a dog of this description, and had been so often reminded of it that he had associated the sound with the incident.

I know a little dog in New York that understands the same sound in a similar way and for similar reasons. She also recognises the name of the lady who owns the black dog. A family, with whom I am on close terms of friendship, owns an ugly little mongrel, to which two of the daughters are very devoted. They have reared her with great care, and lavished upon her many luxuries, far better than most human beings enjoy. The young ladies declared to me that Eunice (which is the dog's name) could understand every word they said on any subject that she had been accustomed to hear.

Mattie would say to her, "Eunice, go tell Miss Kate to get on her hat and let us go take a walk." The little dog would run to Miss Kate's room and bark and jump until the young lady would comply. I found that the dog associated the sounds "hat" and "walk" with the act of taking a stroll in the company of the young ladies; but she would act just the same when either one of these words were said to her as she would if one were to repeat a whole canto of Milton; and I think the young ladies have never quite forgiven me for trying to prove to them that Eunice was not a fine English scholar.

I find, by means of many experiments, that much depends upon the manner of delivering these sounds; but that the animal is largely guided by the sound alone is proven by the fact that some dogs understand English, others French, German, or some other language, and they do not really understand unless addressed in the speech with which they are familiar.

A short time since I tried a novel experiment with the phonograph and two black-and-tan terriers, mother and son. The son was a notorious talker in the way of barking almost continuously at everything, and on all occasions and at all times, while the mother was naturally taciturn, good-natured, and fairly intelligent. I first took the son to a room where I had the phonograph, and I made a record of a number of sounds of his voice. The children aided me in the experiment by getting him to talk for food, bark at his image in the mirror, and by various other ways they induced him to other sounds in the presence of the phonograph. A few days later I had them bring the mother to the same place, where I discharged the contents of my phonograph cylinder in her presence. She gave every evidence of recognising the sounds of the young dog, and in a few instances responded to them. She was naturally perplexed at not being able to find him, and searched the horn and various parts of the room in quest of the young dog. I delivered to her at the same time the record of another dog, to which she paid little attention except by an occasional growl and a look into the horn to see what it meant. She evidently recognised the sounds of the young dog with which she was familiar and seemed to interpret their meanings, whereas the sounds from the other cylinder did little more than attract her attention.

Last summer I stopped at a small town in Northern Virginia. A young man at the same hotel had two setters and a black-and-tan terrier. I experimented extensively with these three dogs during my stay, and deduced therefrom some conclusions which were inevitable. The hotel verandah opened on the street, and was a place of resort for gentlemen of leisure about town. There was also a side entrance through a large yard. I have frequently observed the dogs lying asleep on the verandah, when the owner would enter the side yard on a flagstone walk, often in the midst of conversation of a dozen men. The terrier would recognise the footsteps of his master, would utter a low sound and spring to his feet, and rush at once in the direction whence he heard the steps. The setters invariably seemed to know what it meant, would raise their heads, lash their tails upon the floor, showing evident signs of understanding the situation. I have seen this terrier recognise the steps of his master when the latter was accompanied by two or three other persons. The delicate precision of his hearing was marvellous, and in no instance, so far as I observed, was he deceived in the approaching footsteps. I cannot believe that he was guided by the sense of smell, as it is evident that the setters, whose habits of hunting have developed in them a much more sensitive olfactory power, would naturally have been the first to have detected their master's approach, and yet it was equally evident that the terrier's ears were the first to catch the sounds.

I have observed among dogs associated with each other that where one should bark in the distance, as though he had something at bay, his companion, hearing him from the house, would prick up his ears, listen for a moment, and then dash off in the direction from whence the sounds came; whereas the bark of a strange dog, even having something at bay, would only cause him to listen, utter a low sound or grunt, and lie down again and take a nap, as much as to say "That's nothing to me!" I have known many instances where dogs would follow the farm waggon to town, and faithfully guard the waggon and its contents all day long, with a fidelity that we seldom see in the most devoted servants. The attachment of a dog to his master has been a subject of remark from time immemorial, until the saying has crystallised into a maxim--"As faithful as a watch-dog." A friend of mine had a little terrier, whose name was Nicodemus, that had a habit of sitting in the kitchen window to watch people pass the street. She assures me that on washdays, when the steam condensed on the window-panes, Nicodemus would lick the moisture from the glass in order to see through it more clearly. Could instinct be the guide in such an act?

If man would only pause and calmly view the facts, he would find that he is but a joint heir of Nature; and why not so? From a religious point of view I cannot doubt that the wisdom and mercy of God would bestow alike on all the faculties of speech and reason as their conditions of life require them, and from a scientific point of view I cannot charge the laws of evolution with such disorder. In either case it were a harsh and jarring discord in the great harp of Nature, whether played by the hand of Chance or swept by the fingers of Omniscience.

* * * * *